A THOUSAND BRAINS

A THOUSAND BRAINS

A NEW THEORY OF INTELLIGENCE

JEFF HAWKINS

With a foreword by Richard Dawkins

BASIC BOOKS
New York

Basic Books
Hachette Book Group
1290 Avenue of the Americas, New York, NY 10104
www.basicbooks.com

Printed in the United States of America

First Edition: March 2021

Published by Basic Books, an imprint of Perseus Books, LLC, a subsidiary of Hachette Book Group, Inc. The Basic Books name and logo is a trademark of the Hachette Book Group.

The Hachette Speakers Bureau provides a wide range of authors for speaking events. To find out more, go to www.hachettespeakersbureau.com or call (866) 376-6591.

The publisher is not responsible for websites (or their content) that are not owned by the publisher.

Print book interior design by Trish Wilkinson

Library of Congress Cataloging-in-Publication Data

Names: Hawkins, Jeff, 1957– author.
Title: A thousand brains : a new theory of intelligence / Jeff Hawkins ; with a foreword by Richard Dawkins.
Description: First edition. | New York : Basic Books, 2021. | Includes bibliographical references and index.
Identifiers: LCCN 2020038829 | ISBN 9781541675810 (hardcover) | ISBN 9781541675803 (ebook)
Subjects: LCSH: Brain. | Intellect. | Artificial intelligence.
Classification: LCC QP376 .H2944 2021 | DDC 612.8/2—dc23
LC record available at https://lccn.loc.gov/2020038829

ISBNs: 978-1-5416-7581-0 (hardcover), 978-1-5416-7580-3 (ebook)

LSC-H

Printing 8, 2022

Contents

Foreword by Richard Dawkins

Don't read this book at bedtime. Not that it's frightening. It won't give you nightmares. But it is so exhilarating, so stimulating, it'll turn your mind into a whirling maelstrom of excitingly provocative ideas—you'll want to rush out and tell someone rather than go to sleep. It is a victim of this maelstrom who writes the foreword, and I expect it'll show.

Charles Darwin was unusual among scientists in having the means to work outside universities and without government research grants. Jeff Hawkins might not relish being called the Silicon Valley equivalent of a gentleman scientist but—well, you get the parallel. Darwin's powerful idea was too revolutionary to catch on when expressed as a brief article, and the Darwin-Wallace joint papers of 1858 were all but ignored. As Darwin himself said, the idea needed to be expressed at book length. Sure enough, it was his great book that shook Victorian foundations, a year later. Book-length treatment, too, is needed for Jeff Hawkins's Thousand Brains Theory. And for his notion of reference frames—"The very act of thinking is a form of movement"—bull's-eye! These two ideas are each profound enough to fill a book. But that's not all.

T. H. Huxley famously said, on closing *On the Origin of Species*, "How extremely stupid of me not to have thought of that." I'm not suggesting that brain scientists will necessarily say the same when they close this book. It is a book of many exciting ideas, rather than one huge idea like Darwin's.

I suspect that not just T. H. Huxley but his three brilliant grandsons would have loved it: Andrew because he discovered how the nerve impulse works (Hodgkin and Huxley are the Watson and Crick of the nervous system); Aldous because of his visionary and poetic voyages to the mind's furthest reaches; and Julian because he wrote this poem, extolling the brain's capacity to construct a model of reality, a microcosm of the universe:

> *The world of things entered your infant mind*
> *To populate that crystal cabinet.*
> *Within its walls the strangest partners met,*
> *And things turned thoughts did propagate their kind.*
>
> *For, once within, corporeal fact could find*
> *A spirit. Fact and you in mutual debt*
> *Built there your little microcosm—which yet*
> *Had hugest tasks to its small self assigned.*
>
> *Dead men can live there, and converse with stars:*
> *Equator speaks with pole, and night with day;*
> *Spirit dissolves the world's material bars—*
> *A million isolations burn away.*
> *The Universe can live and work and plan,*
> *At last made God within the mind of man.*

The brain sits in darkness, apprehending the outside world only through a hailstorm of Andrew Huxley's nerve impulses. A nerve impulse from the eye is no different from one from the ear or the big toe. It's where they end up in the brain that sorts them out. Jeff

Hawkins is not the first scientist or philosopher to suggest that the reality we perceive is a constructed reality, a model, updated and informed by bulletins streaming in from the senses. But Hawkins is, I think, the first to give eloquent space to the idea that there is not one such model but thousands, one in each of the many neatly stacked columns that constitute the brain's cortex. There are about 150,000 of these columns and they are the stars of the first section of the book, along with what he calls "frames of reference." Hawkins's thesis about both of these is provocative, and it'll be interesting to see how it is received by other brain scientists: well, I suspect. Not the least fascinating of his ideas here is that the cortical columns, in their world-modeling activities, work semi-autonomously. What "we" perceive is a kind of democratic consensus from among them.

Democracy in the brain? Consensus, and even dispute? What an amazing idea. It is a major theme of the book. We human mammals are the victims of a recurrent dispute: a tussle between the old reptilian brain, which unconsciously runs the survival machine, and the mammalian neocortex sitting in a kind of driver's seat atop it. This new mammalian brain—the cerebral cortex—thinks. It is the seat of consciousness. It is aware of past, present, and future, and it sends instructions to the old brain, which executes them.

The old brain, schooled by natural selection over millions of years when sugar was scarce and valuable for survival, says, "Cake. Want cake. Mmmm cake. Gimme." The new brain, schooled by books and doctors over mere tens of years when sugar was overplentiful, says, "No, no. Not cake. Mustn't. Please don't eat that cake." Old brain says, "Pain, pain, horrible pain, stop the pain *immediately*." New brain says, "No, no, bear the torture, don't betray your country by surrendering to it. Loyalty to country and comrades comes before even your own life."

The conflict between the old reptilian and the new mammalian brain furnishes the answer to such riddles as "Why does pain have to be so damn painful?" What, after all, is pain for? Pain is a proxy for death. It is a warning to the brain, "Don't do that again: don't tease a snake, pick up a hot ember, jump from a great height. This time it only hurt; next time it might kill you." But now a designing engineer might say what we need here is the equivalent of a painless flag in the brain. When the flag shoots up, don't repeat whatever you just did. But instead of the engineer's easy and painless flag, what we actually get is pain—often excruciating, unbearable pain. Why? What's wrong with the sensible flag?

The answer probably lies in the disputatious nature of the brain's decision-making processes: the tussle between old brain and new brain. It being too easy for the new brain to overrule the vote of the old brain, the painless flag system wouldn't work. Neither would torture.

The new brain would feel free to ignore my hypothetical flag and endure any number of bee stings or sprained ankles or torturers' thumbscrews if, for some reason, it "wanted to." The old brain, which really "cares" about surviving to pass on the genes, might "protest" in vain. Maybe natural selection, in the interests of survival, has ensured "victory" for the old brain by making pain so damn painful that the new brain cannot overrule it. As another example, if the old brain were "aware" of the betrayal of sex's Darwinian purpose, the act of donning a condom would be unbearably painful.

Hawkins is on the side of the majority of informed scientists and philosophers who will have no truck with dualism: there is no ghost in the machine, no spooky soul so detached from hardware that it survives the hardware's death, no Cartesian theatre (Dan Dennett's term) where a colour screen displays a movie of the world to a watching self. Instead, Hawkins proposes multiple models of the world, constructed microcosms, informed and adjusted by the rain of nerve impulses pouring in from the senses. By the way, Hawkins doesn't totally rule out the long-term future possibility of escaping death by uploading your brain to a computer, but he doesn't think it would be much fun.

Among the more important of the brain's models are models of the body itself, coping, as they must, with how the body's own movement changes our perspective on the world outside the prison wall of the skull. And this is relevant to the major preoccupation of the middle section of the book, the intelligence of machines. Jeff Hawkins has great respect, as do I, for those smart people, friends of his and mine, who fear the approach of superintelligent machines to supersede us, subjugate us, or even dispose of us altogether. But Hawkins doesn't fear them, partly because the faculties that make for mastery of chess or Go are not those that can cope with the complexity of the real world. Children who can't play chess "know how liquids spill, balls roll, and dogs bark. They know how to use pencils, markers, paper, and glue. They know how to open books and that paper can rip." And they have a self-image, a body image that emplaces them in the world of physical reality and allows them to navigate effortlessly through it.

It is not that Hawkins underestimates the power of artificial intelligence and the robots of the future. On the contrary. But he thinks most present-day research is going about it the wrong way. The right way, in his view, is to understand how the brain works and to borrow its ways but hugely speed them up.

And there is no reason to (indeed, please let's not) borrow the ways of the old brain, its lusts and hungers, cravings and angers, feelings and fears, which can drive us along paths seen as harmful by the new brain. Harmful at least from the perspective that Hawkins and I, and almost certainly you, value. For he is very clear that our enlightened values must, and do, diverge sharply from the primary and primitive value of our selfish genes—the raw imperative to reproduce at all costs. Without an old brain, in his view (which I suspect may be controversial), there is no reason to expect an AI to harbour malevolent feelings toward us. By the same token, and also perhaps controversially, he doesn't think switching off a conscious AI would be murder: Without an old brain, why would it feel fear or sadness? Why would it want to survive?

In the chapter "Genes Versus Knowledge," we are left in no doubt about the disparity between the goals of old brain (serving selfish genes) and of the new brain (knowledge). It is the glory of the human cerebral cortex that it—unique among all animals and unprecedented in all geological time—has the power to defy the dictates of the selfish genes. We can enjoy sex without procreation. We can devote our lives to philosophy, mathematics, poetry, astrophysics, music, geology, or the warmth of human love, in defiance of the old brain's genetic urging that these are a waste of time— time that "should" be spent fighting rivals and pursuing multiple sexual partners: "As I see it, we have a profound choice to make. It is a choice between favoring the old brain or favoring the new brain. More specifically, do we want our future to be driven by the processes that got us here, namely, natural selection, competition, and the drive of selfish genes? Or, do we want our future to be driven by intelligence and its desire to understand the world?"

I began by quoting T. H. Huxley's endearingly humble remark on closing Darwin's *Origin*. I'll end with just one of Jeff Hawkins's many fascinating ideas—he wraps it up in a mere couple of pages—which had me echoing Huxley. Feeling the need for a cosmic tombstone, something to let the galaxy know that we were once here and capable of announcing the fact, Hawkins notes that all civilisations are ephemeral. On the scale of universal time, the interval between a civilisation's invention of electromagnetic communication and its extinction is like the flash of a firefly. The chance of any one flash coinciding with another is unhappily small. What we need, then—why I called it a tombstone—is a message that says not "We are here" but "We were once here." And the tombstone must have cosmic-scale duration: not only must it be visible from parsecs away, it must last for millions if not billions of years, so that it is still proclaiming its message when other flashes of intellect intercept it long after our extinction. Broadcasting prime numbers or the digits of π won't cut it. Not as a radio signal or a pulsed laser beam, anyway. They certainly proclaim biological intelligence, which is why they are the stock-in-trade of SETI (the search for extraterrestrial intelligence) and science fiction, but they are too brief, too in the present. So, what signal would last long enough and be detectable from a very great distance in any direction? This is where Hawkins provoked my inner Huxley.

It's beyond us today, but in the future, before our firefly flash is spent, we could put into orbit around the Sun a series of satellites "that block a bit of the Sun's light in a pattern that would not occur naturally. These orbiting Sun blockers would continue to orbit the Sun for millions of years, long after we are gone, and they could be detected from far away." Even if the spacing of these umbral satellites is not literally a series of prime numbers, the message could be made unmistakable: "Intelligent Life Woz 'Ere."

What I find rather pleasing—and I offer the vignette to Jeff Hawkins to thank him for the pleasure his brilliant book has given me—is that a cosmic message coded in the form of a pattern of intervals between spikes (or in his case anti-spikes, as his satellites dim the Sun) would be using the same kind of code as a neuron.

This is a book about how the brain works. It works the brain in a way that is nothing short of exhilarating.

PART 1

A New Understanding of the Brain

The cells in your head are reading these words. Think of how remarkable that is. Cells are simple. A single cell can't read, or think, or do much of anything. Yet, if we put enough cells together to make a brain, they not only read books, they write them. They design buildings, invent technologies, and decipher the mysteries of the universe. How a brain made of simple cells creates intelligence is a profoundly interesting question, and it remains a mystery.

Understanding how the brain works is considered one of humanity's grand challenges. The quest has spawned dozens of national and international initiatives, such as Europe's Human Brain Project and the International Brain Initiative. Tens of thousands of neuroscientists work in dozens of specialties, in practically every country in the world, trying to understand the brain. Although neuroscientists study the brains of different animals and ask varied questions, the ultimate goal of neuroscience is to learn how the human brain gives rise to human intelligence.

You might be surprised by my claim that the human brain remains a mystery. Every year, new brain-related discoveries are announced, new brain books are published, and researchers in related fields such as artificial intelligence claim their creations are approaching the intelligence of, say, a mouse or a cat. It would be easy to conclude from this that scientists have a pretty good idea of how the brain works. But if you ask neuroscientists, almost all of them would admit that we are still in the dark. We have learned a tremendous amount of knowledge and facts about the brain, but we have little understanding of how the whole thing works.

In 1979, Francis Crick, famous for his work on DNA, wrote an essay about the state of brain science, titled "Thinking About the Brain." He described the large quantity of facts that scientists had collected about the brain, yet, he concluded, "in spite of the steady accumulation of detailed knowledge, how the human brain works is still profoundly mysterious." He went on to say, "What is conspicuously lacking is a broad framework of ideas in which to interpret these results."

Crick observed that scientists had been collecting data on the brain for decades. They knew a great many facts. But no one had figured out how to assemble those facts into something meaningful. The brain was like a giant jigsaw puzzle with thousands of pieces. The puzzle pieces were sitting in front of us, but we could not make sense of them. No one knew what the solution was supposed to look like. According to Crick, the brain was a mystery not because we hadn't collected enough data, but because we didn't know how to arrange the pieces we already had. In the forty years since Crick wrote his essay there have been many significant discoveries about the brain, several of which I will talk about later, but overall his observation is still true. How intelligence arises from cells in your head is still a profound mystery. As more puzzle pieces are collected each year, it sometimes feels as if we are getting further from understanding the brain, not closer.

I read Crick's essay when I was young, and it inspired me. I felt that we could solve the mystery of the brain in my lifetime, and I have pursued that goal ever since. For the past fifteen years, I have led a research team in Silicon Valley that studies a part of the brain called the neocortex. The neocortex occupies about 70 percent of the volume of a human brain and it is responsible for everything we associate with intelligence, from our senses of vision, touch, and hearing, to language in all its forms, to abstract thinking such as mathematics and philosophy. The goal of our research is to understand how the neocortex works in sufficient detail that we can explain the biology of the brain and build intelligent machines that work on the same principles.

In early 2016 the progress of our research changed dramatically. We had a breakthrough in our understanding. We realized that we and other scientists had missed a key ingredient. With this new insight, we saw how the pieces of the puzzle fit together. In other words, I believe we discovered the framework that Crick wrote about, a framework that not only explains the basics of how the neocortex works but also gives rise to a new way to think about intelligence. We do not yet have a complete theory of the brain—far from it. Scientific fields typically start with a theoretical framework and only later do the details get worked out. Perhaps the most famous example is Darwin's theory of evolution. Darwin proposed a bold new way of thinking about the origin of species, but the details, such as how genes and DNA work, would not be known until many years later.

To be intelligent, the brain has to learn a great many things about the world. I am not just referring to what we learn in school, but to basic things, such as what everyday objects look, sound, and feel like. We have to learn how objects behave, from how doors open and close to what the apps on our smartphones do when we touch the screen. We need to learn where everything is located in the world, from where you keep your personal possessions in your home to where the library and post office are in your town. And of course, we learn higher-level concepts, such as the meaning of "compassion" and "government." On top of all this, each of us learns the meaning of tens of thousands of words. Every one of us possesses a tremendous amount of knowledge about the world. Some of our basic skills are determined by our genes, such as how to eat or how to recoil from pain. But most of what we know about the world is learned.

Scientists say that the brain learns a model of the world. The word "model" implies that what we know is not just stored as a pile of facts but is organized in a way that reflects the structure of the world and everything it contains. For example, to know what a bicycle is, we don't remember a list of facts about bicycles. Instead, our brain creates a model of bicycles that includes the different parts, how the parts are arranged relative to each other, and how the different parts move and work together. To recognize something, we need to first learn what it looks and feels like, and to achieve goals we need to learn how things in the world typically behave when we interact with them. Intelligence is intimately tied to the brain's model of the world; therefore, to understand how the brain creates intelligence, we have to figure out how the brain, made of simple cells, learns a model of the world and everything in it.

Our 2016 discovery explains how the brain learns this model. We deduced that the neocortex stores everything we know, all our knowledge, using something called reference frames. I will explain this more fully later, but for now, consider a paper map as an analogy. A map is a type of model: a map of a town is a model of the town, and the grid lines, such as lines of latitude and longitude, are a type of reference frame. A map's grid lines, its reference frame, provide the structure of the map. A reference frame tells you where things are located relative to each other, and it can tell you how to achieve goals, such as how to get from one location to another. We realized that the brain's model of the world is built using maplike reference frames. Not one reference frame, but hundreds of thousands of them. Indeed, we now understand that most of the cells in your neocortex are dedicated to creating and manipulating reference frames, which the brain uses to plan and think.

With this new insight, answers to some of neuroscience's biggest questions started to come into view. Questions such as, How do our varied sensory inputs get united into a singular experience? What is happening when we think? How can two people reach different beliefs from the same observations? And why do we have a sense of self?

This book tells the story of these discoveries and the implications they have for our future. Most of the material has been published in scientific journals. I provide links to these papers at the end of the book. However, scientific papers are not well suited for explaining large-scale theories, especially in a way that a nonspecialist can understand.

I have divided the book into three parts. In the first part, I describe our theory of reference frames, which we call the Thousand Brains Theory. The theory is partly based on logical deduction, so I will take you through the steps we took to reach our conclusions. I will also give you a bit of historical background to help you see how the theory relates to the history of thinking about the brain. By the end of the first part of the book, I hope you will have an understanding of what is going on in your head as you think and act within the world, and what it means to be intelligent.

The second part of the book is about machine intelligence. The twenty-first century will be transformed by intelligent machines in the same way that the twentieth century was transformed by computers. The Thousand Brains Theory explains why today's AI is not yet intelligent and what we need to do to make truly intelligent machines. I describe what intelligent machines in the future will look like and how we might use them. I explain why some machines will be conscious and what, if anything, we should do about it. Finally, many people are worried that intelligent machines are an existential risk, that we are about to create a technology that will destroy humanity. I disagree. Our discoveries illustrate why machine intelligence, on its own, is benign. But, as a powerful technology, the risk lies in the ways humans might use it.

In the third part of the book, I look at the human condition from the perspective of the brain and intelligence. The brain's model of the world includes a model of our self. This leads to the strange truth that what you and I perceive, moment to moment, is a simulation of the world, not the real world. One consequence of the Thousand Brains Theory is that our beliefs about the world can be false. I explain how this can occur, why false beliefs can be difficult to eliminate, and how false beliefs combined with our more primitive emotions are a threat to our long-term survival.

The final chapters discuss what I consider to be the most important choice we will face as a species. There are two ways to think about ourselves. One is as biological organisms, products of evolution and natural selection. From this point of view, humans are defined by our genes, and the purpose of life is to replicate them. But we are now emerging from our purely biological past. We have become an intelligent species. We are the first species on Earth to know the size and age of the universe. We are the first species to know how the Earth evolved and how we came to be. We are the first species to develop tools that allow us to explore the universe and learn its secrets. From this point of view, humans are defined by our intelligence and our knowledge, not by our genes. The choice we face as we think about the future is, should we continue to be driven by our biological past or choose instead to embrace our newly emerged intelligence?

We may not be able to do both. We are creating powerful technologies that can fundamentally alter our planet, manipulate biology, and soon, create machines that are smarter than we are. But we still possess the primitive behaviors that got us to this point. This combination is the true existential risk that we must address. If we are willing to embrace intelligence and knowledge as what defines us, instead of our genes, then perhaps we can create a future that is longer lasting and has a more noble purpose.

The journey that led to the Thousand Brains Theory has been long and convoluted. I studied electrical engineering in college and had just started my first job at Intel when I read Francis Crick's essay. It had such a profound effect on me that I decided to switch careers and dedicate my life to studying the brain. After an unsuccessful attempt to get a position studying brains at Intel, I applied to be a graduate student at MIT's AI lab. (I felt that the best way to build intelligent machines was first to study the brain.) In my interviews with MIT faculty, my proposal to create intelligent machines based on brain theory was rejected. I was told that the brain was just a messy computer and there was no point in studying it. Crestfallen but undeterred, I next enrolled in a neuroscience PhD program at the University of California, Berkeley. I started my studies in January 1986.

Upon arriving at Berkeley, I reached out to the chair of the graduate group of neurobiology, Dr. Frank Werblin, for advice. He asked me to write a paper describing the research I wanted to do for my PhD thesis. In the paper, I explained that I wanted to work on a theory of the neocortex. I knew that I wanted to approach the problem by studying how the neocortex makes predictions. Professor Werblin had several faculty members read my paper, and it was well received. He told me that my ambitions were admirable, my approach was sound, and the problem I wanted to work on was one of the most important in science, but—and I didn't see this coming—he didn't see how I could pursue my dream at that time. As a neuroscience graduate student, I would have to work for a professor, doing similar work to what the professor was already working on. And no one at Berkeley, or anywhere else that he knew of, was doing something close enough to what I wanted to do.

Trying to develop an overall theory of brain function was considered too ambitious and therefore too risky. If a student worked on this for five years and didn't make progress, they might not graduate. It was similarly risky for professors; they might not get tenure. The agencies that dispensed funding for research also thought it was too risky. Research proposals that focused on theory were routinely rejected.

I could have worked in an experimental lab, but after interviewing at a few I knew that it wasn't a good fit for me. I would be spending most of my hours training animals, building experimental equipment, and collecting data. Any theories I developed would be limited to the part of the brain studied in that lab.

For the next two years, I spent my days in the university's libraries reading neuroscience paper after neuroscience paper. I read hundreds of them, including all the most important papers published over the previous fifty years. I also read what psychologists, linguists, mathematicians, and philosophers thought about the brain and intelligence. I got a first-class, albeit unconventional, education. After two years of self-study, a change was needed. I came up with a plan. I would work again in industry for four years and then reassess my opportunities in academia. So, I went back to working on personal computers in Silicon Valley.

I started having success as an entrepreneur. From 1988 to 1992, I created one of the first tablet computers, the GridPad. Then in 1992, I founded Palm Computing, beginning a ten-year span when I designed some of the first handheld computers and smartphones such as the PalmPilot and the Treo. Everyone who worked with me at Palm knew that my heart was in neuroscience, that I viewed my work in mobile computing as temporary. Designing some of the first handheld computers and smartphones was exciting work. I knew that billions of people would ultimately rely on these devices, but I felt that understanding the brain was even more important. I believed that brain theory would have a bigger positive impact on the future of humanity than computing. Therefore, I needed to return to brain research.

There was no convenient time to leave, so I picked a date and walked away from the businesses I helped create. With the assistance and prodding of a few neuroscientist friends, (notably Bob Knight at UC Berkeley, Bruno Olshausen at UC Davis, and Steve Zornetzer at NASA Ames Research), I created the Redwood Neuroscience Institute (RNI) in 2002. RNI focused exclusively on neocortical theory and had ten full-time scientists. We were all interested in large-scale theories of the brain, and RNI was one of the only places in the world where this focus was not only tolerated but expected. Over the course of the three years that I ran RNI, we had over one hundred visiting scholars, some of whom stayed for days or weeks. We had weekly lectures, open to the public, which usually turned into hours of discussion and debate.

Everyone who worked at RNI, including me, thought it was great. I got to know and spend time with many of the world's top neuroscientists. It allowed me to become knowledgeable in multiple fields of neuroscience, which is difficult to do with a typical academic position. The problem was that I wanted to know the answers to a set of specific questions, and I didn't see the team moving toward consensus on those questions. The individual scientists were content to do their own thing. So, after three years of running an institute, I decided the best way to achieve my goals was to lead my own research team.

RNI was in all other ways doing well, so we decided to move it to UC Berkeley. Yes, the same place that told me I couldn't study brain theory decided, nineteen years later, that a brain-theory center was exactly what they needed. RNI continues today as the Redwood Center for Theoretical Neuroscience.

As RNI moved to UC Berkeley, several colleagues and I started Numenta. Numenta is an independent research company. Our primary goal is to develop a theory of how the neocortex works. Our secondary goal is to apply what we learn about brains to machine learning and machine intelligence. Numenta is similar to a typical research lab at a university, but with more flexibility. It allows me to direct a team, make sure we are all focused on the same task, and try new ideas as often as needed.

As I write, Numenta is over fifteen years old, yet in some ways we are still like a start-up. Trying to figure out how the neocortex works is extremely challenging. To make progress, we need the flexibility and focus of a start-up environment. We also need a lot of patience, which is not typical for a start-up. Our first significant discovery—how neurons make predictions—occurred in

2010, five years after we started. The discovery of maplike reference frames in the neocortex occurred six years later in 2016.

In 2019, we started to work on our second mission, applying brain principles to machine learning. That year was also when I started writing this book, to share what we have learned.

I find it amazing that the only thing in the universe that knows the universe exists is the three-pound mass of cells floating in our heads. It reminds me of the old puzzle: If a tree falls in the forest and no one is there to hear it, did it make a sound? Similarly, we can ask: If the universe came into and out of existence and there were no brains to know it, did the universe really exist? Who would know? A few billion cells suspended in your skull know not only that the universe exists but that it is vast and old. These cells have learned a model of the world, knowledge that, as far as we know, exists nowhere else. I have spent a lifetime striving to understand how the brain does this, and I am excited by what we have learned. I hope you are excited as well. Let's get started.

CHAPTER 1

Old Brain—New Brain

To understand how the brain creates intelligence, there are a few basics you need to know first.

Shortly after Charles Darwin published his theory of evolution, biologists realized that the human brain itself had evolved over time, and that its evolutionary history is evident from just looking at it. Unlike species which often disappear as new ones appear, the brain evolved by adding new parts on top of the older parts. For example, some of the oldest and simplest nervous systems are sets of neurons that run down the back of tiny worms. These neurons allow the worm to make simple movements, and they are the predecessor of our spinal cord, which is similarly responsible for many of our basic movements. Next to appear was a lump of neurons at one end of the body that controlled functions such as digestion and breathing. This lump is the predecessor of our brain stem, which similarly controls our digestion and breathing. The brain stem extended what was already there, but it did not replace it. Over time, the brain grew capable of increasingly complex behaviors by evolving new parts on top of the older parts. This method of growth by addition applies to the brains of most complex animals. It is easy to see why the old brain parts are still there.

No matter how smart and sophisticated we are, breathing, eating, sex, and reflex reactions are still critical to our survival.

The newest part of our brain is the neocortex, which means "new outer layer." All mammals, and only mammals, have a neocortex. The human neocortex is particularly large, occupying about 70 percent of the volume of our brain. If you could remove the neocortex from your head and iron it flat, it would be about the size of a large dinner napkin and twice as thick (about 2.5 millimeters). It wraps around the older parts of the brain such that when you look at a human brain, most of what you see is the neocortex (with its characteristic folds and creases), with bits of the old brain and the spinal cord sticking out the bottom.

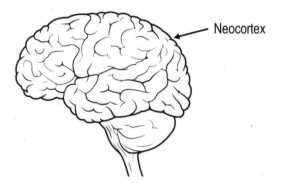

Neocortex

A human brain

The neocortex is the organ of intelligence. Almost all the capabilities we think of as intelligence—such as vision, language, music, math, science, and engineering—are created by the neocortex. When we think about something, it is mostly the neocortex doing the thinking. Your neocortex is reading or listening to this book, and my neocortex is writing this book. If we want to understand intelligence, then we have to understand what the neocortex does and how it does it.

An animal doesn't need a neocortex to live a complex life. A crocodile's brain is roughly equivalent to our brain, but without a proper neocortex. A crocodile has sophisticated behaviors, cares for its young, and knows how to navigate its environment. Most people would say a crocodile has some level of intelligence, but nothing close to human intelligence.

The neocortex and the older parts of the brain are connected via nerve fibers; therefore, we cannot think of them as completely separate organs. They are more like roommates, with separate agendas and personalities, but who need to cooperate to get anything done. The neocortex is in a decidedly unfair position, as it doesn't control behavior directly. Unlike other parts of the brain, none of the cells in the neocortex connect directly to muscles, so it can't, on its own, make any muscles move. When the neocortex wants to do something, it sends a signal to the old brain, in a sense asking the old brain to do its bidding. For example, breathing is a function of the brain stem, requiring no thought or input from the neocortex. The neocortex can temporarily control breathing, as when you consciously decide to hold your breath. But if the brain stem detects that your body needs more oxygen, it will ignore the neocortex and take back control. Similarly, the neocortex might think, "Don't eat this piece of cake. It isn't healthy." But if older and more primitive parts of the brain say, "Looks good, smells good, eat it," the cake can be hard to resist. This battle between the old and new brain is an underlying theme of this book. It will play an important role when we discuss the existential risks facing humanity.

The old brain contains dozens of separate organs, each with a specific function. They are visually distinct, and their shapes, sizes, and connections reflect what they do. For example, there are several pea-size organs in the amygdala, an older part of the brain, that are responsible for different types of aggression, such as premeditated and impulsive aggression.

The neocortex is surprisingly different. Although it occupies almost three-quarters of the brain's volume and is responsible for a myriad of cognitive functions, it has no visually obvious divisions. The folds and creases are needed to fit the neocortex into the skull, similar to what you would see if you forced a napkin into a large wine glass. If you ignore the folds and creases, then the neocortex looks like one large sheet of cells, with no obvious divisions.

Nonetheless, the neocortex is still divided into several dozen areas, or regions, that perform different functions. Some of the regions are responsible for vision, some for hearing, and some for touch. There are regions responsible for language and planning. When the neocortex is damaged, the deficits that arise depend on what part of the neocortex is affected. Damage to the back of the head results in blindness, and damage to the left side could lead to loss of language.

The regions of the neocortex connect to each other via bundles of nerve fibers that travel under the neocortex, the so-called white matter of the brain. By carefully following these nerve fibers, scientists can determine how many regions there are and how they are connected. It is difficult to study human brains, so the first complex mammal that was analyzed this way was the macaque monkey. In 1991, two scientists, Daniel Felleman and David Van Essen, combined data from dozens of separate studies to create a famous illustration of the macaque monkey's neocortex. Here is one of the images that they created (a map of a human's neocortex would be different in detail, but similar in overall structure).

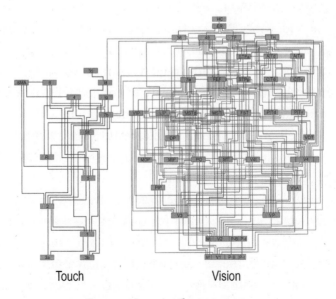

Touch   Vision

Connections in the neocortex

The dozens of small rectangles in this picture represent the different regions of the neocortex, and the lines represent how information flows from one region to another via the white matter.

A common interpretation of this image is that the neocortex is hierarchical, like a flowchart. Input from the senses enters at the bottom (in this diagram, input from the skin is on the left and input from the eyes is on the right). The input is processed in a series of steps, each of which extracts more and more complex features from the input. For example, the first region that gets input from the eyes might detect simple patterns such as lines or edges. This information is sent to the next region, which might detect more complex features such as corners or shapes. This stepwise process continues until some regions detect complete objects.

There is a lot of evidence supporting the flowchart hierarchy interpretation. For example, when scientists look at cells in regions at the bottom of the hierarchy, they see that they respond best to simple features, while cells in the next region respond to more complex features. And sometimes they find cells in higher regions that respond to complete objects. However, there is also a lot of evidence that suggests the neocortex is not like a flowchart. As you can see in the diagram, the regions aren't arranged one on top of another as they would be in a flowchart. There are multiple regions at each level, and most regions connect to multiple levels of the hierarchy. In fact, the majority of connections between regions do not fit into a hierarchical scheme at all. In addition, only some of the cells in each region act like feature detectors; scientists have not been able to determine what the majority of cells in each region are doing.

We are left with a puzzle. The organ of intelligence, the neocortex, is divided into dozens of regions that do different things, but on the surface, they all look the same. The regions connect in a complex mishmash that is somewhat like a flowchart but mostly not. It is not immediately clear why the organ of intelligence looks this way.

The next obvious thing to do is to look inside the neocortex, to see the detailed circuitry within its 2.5 mm thickness. You might imagine that, even if different areas of the neocortex look the same on the outside, the detailed neural circuits that create vision, touch, and language would look different on the inside. But this is not the case.

The first person to look at the detailed circuitry inside the neo-cortex was Santiago Ramón y Cajal. In the late 1800s, staining techniques were discovered that allowed individual neurons in the brain to be seen with a microscope. Cajal used these stains to make pictures of every part of the brain. He created thousands of images that, for the first time, showed what brains look like at the cellular level. All of Cajal's beautiful and intricate images of the brain were drawn by hand. He ultimately received the Nobel Prize for his work. Here are two drawings that Cajal made of the neocortex. The one on the left shows only the cell bodies of neurons. The one on the right includes the connections between cells. These images show a slice across the 2.5 mm thickness of the neocortex.

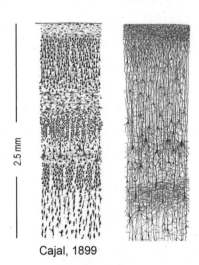

Cajal, 1899

Neurons in a slice of neocortex

The stains used to make these images only color a small per-centage of the cells. This is fortunate, because if every cell were stained then all we would see is black. Keep in mind that the actual number of neurons is much larger than what you see here.

The first thing Cajal and others observed was that the neurons in the neocortex appear to be arranged in layers. The layers, which run parallel to the surface of the neocortex (horizontal in the picture), are caused by differences in the size of the neurons and how closely they are packed. Imagine you had a glass tube and poured in an inch of peas, an inch of lentils, and an inch of soybeans. Looking at the tube from the side you would see three layers. You can see layers in the pictures above. How many layers there are depends on who is doing the counting and the criteria they use for distinguishing the layers. Cajal saw six layers. A simple interpretation is that each layer of neurons is doing something different.

Today we know that there are dozens of different types of neurons in the neocortex, not six. Scientists still use the six-layer terminology. For example, one type of cell might be found in Layer 3 and another in Layer 5. Layer 1 is on the outermost surface of the neocortex closest to the skull, at the top of Cajal's drawing. Layer 6 is closest to the center of the brain, farthest from the skull. It is important to keep in mind that the layers are only a rough guide to where a particular type of neuron might be found. It matters more what a neuron connects to and how it behaves. When you classify neurons by their connectivity, there are dozens of types.

The second observation from these images was that most of the connections between neurons run vertically, between layers. Neurons have treelike appendages called axons and dendrites that allow them to send information to each other. Cajal saw that most of the axons ran between layers, perpendicular to the surface of the neocortex (up and down in the images on page 16). Neurons in some layers make long-distance horizontal connections, but most of the connections are vertical. This means that information arriving in a region of the neocortex moves mostly up and down between the layers before being sent elsewhere.

In the 120 years since Cajal first imaged the brain, hundreds of scientists have studied the neocortex to discover as many details as possible about its neurons and circuits. There are thousands of scientific papers on this topic, far more than I can summarize. Instead, I want to highlight three general observations.

1. The Local Circuits in the Neocortex Are Complex

Under one square millimeter of neocortex (about 2.5 cubic millimeters), there are roughly one hundred thousand neurons, five hundred million connections between neurons (called synapses), and several kilometers of axons and dendrites. Imagine laying out several kilometers of wire along a road and then trying to squish it into two cubic millimeters, roughly the size of a grain of rice. There are dozens of different types of neurons under each square millimeter. Each type of neuron makes prototypical connections to other types of neurons. Scientists often describe regions of the neocortex as performing a simple function, such as detecting features. However, it only takes a handful of neurons to detect features. The precise and extremely complex neural circuits seen everywhere in the neocortex tell us that every region is doing something far more complex than feature detection.

2. The Neocortex Looks Similar Everywhere

The complex circuitry of the neocortex looks remarkably alike in visual regions, language regions, and touch regions. It even looks similar across species such as rats, cats, and humans. There are differences. For example, some regions of the neocortex have more of certain cells and less of others, and there are some regions that have an extra cell type not found elsewhere. Presumably, whatever these regions of the neocortex are doing benefits from these differences. But overall, the variations between regions are relatively small compared to the similarities.

3. Every Part of the Neocortex Generates Movement

For a long time, it was believed that information entered the neo-cortex via the "sensory regions," went up and down the hierarchy of regions, and finally went down to the "motor region." Cells in the motor region project to neurons in the spinal cord that move the muscles and limbs. We now know this description is mislead-ing. In every region they have examined, scientists have found cells that project to some part of the old brain related to movement. For example, the visual regions that get input from the eyes send a signal down to the part of the old brain responsible for moving the eyes. Similarly, the auditory regions that get input from the ears project to the part of the old brain that moves the head. Moving your head changes what you hear, similar to how moving your eyes changes what you see. The evidence we have indicates that the complex circuitry seen everywhere in the neocortex performs a sensory-motor task. There are no pure motor regions and no pure sensory regions.

In summary, the neocortex is the organ of intelligence. It is a napkin-size sheet of neural tissue divided into dozens of regions. There are regions responsible for vision, hearing, touch, and lan-guage. There are regions that are not as easily labeled that are re-sponsible for high-level thought and planning. The regions are connected to each other with bundles of nerve fibers. Some of the connections between the regions are hierarchical, suggesting that information flows from region to region in an orderly fashion like a flowchart. But there are other connections between the regions that seem to have little order, suggesting that information goes all over at once. All regions, no matter what function they perform, look similar in detail to all other regions.

We will meet the first person who made sense of these observa-tions in the next chapter.

This is a good point to say a few words about the style of writing in this book. I am writing for an intellectually curious lay reader. My goal is to convey everything you need to know to understand the new theory, but not a lot more. I assume most readers will have limited prior knowledge of neuroscience. However, if you have a background in neuroscience, you will know where I am omitting details and simplifying complex topics. If that applies to you, I ask for your understanding. There is an annotated reading list at the back of the book where I discuss where to find more details for those who are interested.

CHAPTER 2

Vernon Mountcastle's Big Idea

The Mindful Brain is a small book, only one hundred pages long. Published in 1978, it contains two essays about the brain from two prominent scientists. One of the essays, by Vernon Mountcastle, a neuroscientist at Johns Hopkins University, remains one of the most iconic and important monographs ever written about the brain. Mountcastle proposed a way of thinking about the brain that is elegant—a hallmark of great theories—but also so surprising that it continues to polarize the neuroscience community.

I first read *The Mindful Brain* in 1982. Mountcastle's essay had an immediate and profound effect on me, and, as you will see, his proposal heavily influenced the theory I present in this book.

Mountcastle's writing is precise and erudite but also challenging to read. The title of his essay is the not-too-catchy "An Organizing Principle for Cerebral Function: The Unit Module and the Distributed System." The opening lines are difficult to understand; I include them here so you can get a sense of what his essay feels like.

There can be little doubt of the dominating influence of the Darwinian revolution of the mid-nineteenth century upon concepts of the structure and function of the nervous system.

The ideas of Spencer and Jackson and Sherrington and the many who followed them were rooted in the evolutionary theory that the brain develops in phylogeny by the successive addition of more cephalad parts. On this theory each new addition or enlargement was accompanied by the elaboration of more complex behavior, and at the same time, imposed a regulation upon more caudal and more primitive parts and the presumably more primitive behavior they control.

What Mountcastle says in these first three sentences is that the brain grew large over evolutionary time by adding new brain parts on top of old brain parts. The older parts control more primitive behaviors while the newer parts create more sophisticated ones. Hopefully this sounds familiar, as I discussed this idea in the previous chapter.

However, Mountcastle goes on to say that while much of the brain got bigger by adding new parts on top of old parts, that is not how the neocortex grew to occupy 70 percent of our brain. The neocortex got big by making many copies of the same thing: a basic circuit. Imagine watching a video of our brain evolving. The brain starts small. A new piece appears at one end, then another piece appears on top of that, and then another piece is appended on top of the previous pieces. At some point, millions of years ago, a new piece appears that we now call the neocortex. The neocortex starts small, but then grows larger, not by creating anything new, but by copying a basic circuit over and over. As the neocortex grows, it gets larger in area but not in thickness. Mountcastle argued that, although a human neocortex is much larger than a rat or dog neocortex, they are all made of the same element—we just have more copies of that element.

Mountcastle's essay reminds me of Charles Darwin's book *On the Origin of Species*. Darwin was nervous that his theory of evolution would cause an uproar. So, in the book, he covers a lot of dense and relatively uninteresting material about variation in the animal kingdom before finally describing his theory toward the end. Even then, he never explicitly says that evolution applies to humans. When I read Mountcastle's essay, I get a similar impression. It feels as if Mountcastle knows that his proposal will generate pushback, so he is careful and deliberate in his writing. Here is a second quote from later in Mountcastle's essay:

> *Put shortly, there is nothing intrinsically motor about the motor cortex, nor sensory about the sensory cortex. Thus the elucidation of the mode of operation of the local modular circuit anywhere in the neocortex will be of great generalizing significance.*

In these two sentences, Mountcastle summarizes the major idea put forth in his essay. He says that every part of the neocortex works on the same principle. All the things we think of as intelligence—from seeing, to touching, to language, to high-level thought—are fundamentally the same.

Recall that the neocortex is divided into dozens of regions, each of which performs a different function. If you look at the neocortex from the outside, you can't see the regions; there are no demarcations, just like a satellite image doesn't reveal political borders between countries. If you cut through the neocortex, you see a complex and detailed architecture. However, the details look similar no matter what region of the cortex you cut into. A slice of cortex responsible for vision looks like a slice of cortex responsible for touch, which looks like a slice of cortex responsible for language.

Mountcastle proposed that the reason the regions look similar is that they are all doing the same thing. What makes them different is not their intrinsic function but what they are connected to. If you connect a cortical region to eyes, you get vision; if you connect the same cortical region to ears, you get hearing; and if you connect regions to other regions, you get higher thought, such as language. Mountcastle then points out that if we can discover the basic function of any part of the neocortex, we will understand how the entire thing works.

Mountcastle's idea is as surprising and profound as Darwin's discovery of evolution. Darwin proposed a mechanism—an algorithm, if you will—that explains the incredible diversity of life. What on the surface appears to be many different animals and plants, many types of living things, are in reality manifestations of the same underlying evolutionary algorithm. In turn, Mountcastle is proposing that all the things we associate with intelligence, which on the surface appear to be different, are, in reality, manifestations of the same underlying cortical algorithm. I hope you can appreciate how unexpected and revolutionary Mountcastle's proposal is. Darwin proposed that the diversity of life is due to one basic algorithm. Mountcastle proposed that the diversity of intelligence is also due to one basic algorithm.

Like many things of historical significance, there is some debate as to whether Mountcastle was the first person to propose this idea. It has been my experience that every idea has at least some precedent. But, as far as I know, Mountcastle was the first person to clearly and carefully lay out the argument for a common cortical algorithm.

Mountcastle's and Darwin's proposals differ in one interesting way. Darwin knew *what* the algorithm was: evolution is based on random variation and natural selection. However, Darwin didn't know *where* the algorithm was in the body. This was not known until the discovery of DNA many years later. Mountcastle, by contrast, didn't know *what* the cortical algorithm was; he didn't know what the principles of intelligence were. But he did know *where* this algorithm resided in the brain.

So, what was Mountcastle's proposal for the location of the cortical algorithm? He said that the fundamental unit of the neocortex, the unit of intelligence, was a "cortical column." Looking at the surface of the neocortex, a cortical column occupies about one square millimeter. It extends through the entire 2.5 mm thickness, giving it a volume of 2.5 cubic millimeters. By this definition, there are roughly 150,000 cortical columns stacked side by side in a human neocortex. You can imagine a cortical column is like a little piece of thin spaghetti. A human neocortex is like 150,000 short pieces of spaghetti stacked vertically next to each other.

The width of cortical columns varies from species to species and region to region. For example, in mice and rats, there is one cortical column for each whisker; these columns are about half a millimeter in diameter. In cats, vision columns appear to be about one millimeter in diameter. We don't have a lot of data about the size of columns in a human brain. For simplicity, I will continue to refer to columns as being one square millimeter, endowing each of us with about 150,000 cortical columns. Even though the actual number will likely vary from this, it won't make a difference for our purposes.

Cortical columns are not visible under a microscope. With a few exceptions, there are no visible boundaries between them. Scientists know they exist because all the cells in one column will respond to the same part of the retina, or the same patch of skin, but then cells in the next column will all respond to a different part of the retina or a different patch of skin. This grouping of responses is what defines a column. It is seen everywhere in the neocortex. Mountcastle pointed out that each column is further divided into several hundred "minicolumns." If a cortical column is like a skinny strand of spaghetti, you can visualize minicolumns as even skinnier strands, like individual pieces of hair, stacked side by side inside the spaghetti strand. Each minicolumn contains a little over one hundred neurons spanning all layers. Unlike the larger cortical column, minicolumns are physically distinct and can often be seen with a microscope.

Mountcastle didn't know and he didn't suggest what columns or minicolumns do. He only proposed that each column is doing the same thing and that minicolumns are an important subcomponent.

Let's review. The neocortex is a sheet of tissue about the size of a large napkin. It is divided into dozens of regions that do different things. Each region is divided into thousands of columns. Each column is composed of several hundred hairlike minicolumns, which consist of a little over one hundred cells each. Mountcastle proposed that throughout the neocortex columns and minicolumns performed the same function: implementing a fundamental algorithm that is responsible for every aspect of perception and intelligence.

Mountcastle based his proposal for a universal algorithm on several lines of evidence. First, as I have already mentioned, is that the detailed circuits seen everywhere in the neocortex are remarkably similar. If I showed you two silicon chips with nearly identical circuit designs, it would be safe to assume that they performed nearly identical functions. The same argument applies to the detailed circuits of the neocortex. Second is that the major expansion of the modern human neocortex relative to our hominid ancestors occurred rapidly in evolutionary time, just a few million years. This is probably not enough time for multiple new complex capabilities to be discovered by evolution, but it is plenty of time for evolution to make more copies of the same thing. Third is that the function of neocortical regions is not set in stone. For example, in people with congenital blindness, the visual areas of the neocortex do not get useful information from the eyes. These areas may then assume new roles related to hearing or touch. Finally, there is the argument of extreme flexibility. Humans can do many things for which there was no evolutionary pressure. For example, our brains did not evolve to program computers or make ice cream—both are recent inventions. The fact that we can do these things tells us that the brain relies on a general-purpose method of learning. To me, this last argument is the most compelling. Being able to learn practically anything requires the brain to work on a universal principle.

There are more pieces of evidence that support Mountcastle's proposal. But despite this, his idea was controversial when he introduced it, and it remains somewhat controversial today. I believe there are two related reasons. One is that Mountcastle didn't know what a cortical column does. He made a surprising claim built on a lot of circumstantial evidence, but he didn't propose how a cortical column could actually do all the things we associate with intelligence. The other reason is that the implications of his proposal are hard for some people to believe. For example, you may have trouble accepting that vision and language are fundamentally the same. They don't feel the same. Given these uncertainties, some scientists reject Mountcastle's proposal by pointing out that there are differences between neocortical regions. The differences are relatively small compared to the similarities, but if you focus on them you can argue that different regions of the neocortex are not the same.

Mountcastle's proposal looms in neuroscience like a holy grail. No matter what animal or what part of the brain a neuroscientist studies, somewhere, overtly or covertly, almost all neuroscientists want to understand how the human brain works. And that means understanding how the neocortex works. And that requires understanding what a cortical column does. In the end, our quest to understand the brain, our quest to understand intelligence, boils down to figuring out what a cortical column does and how it does it. Cortical columns are not the only mystery of the brain or the only mystery related to the neocortex. But understanding the cortical column is by far the largest and most important piece of the puzzle.

———————

In 2005, I was invited to give a talk about our research at Johns Hopkins University. I talked about our quest to understand the neocortex, how we were approaching the problem, and the progress we had made. After giving a talk like this, the speaker often meets with individual faculty. On this trip, my final visit was with Vernon Mountcastle and the dean of his department. I felt honored to meet the man who had provided so much insight and inspiration during my life. At one point during our conversation, Mountcastle, who had attended my lecture, said that I should come to work at Johns Hopkins and that he would arrange a position for me. His offer was unexpected and unusual. I could not seriously consider it due to my family and business commitments back in California, but I reflected back to 1986, when my proposal to study the neocortex was rejected by UC Berkeley. How I would have leaped to accept his offer back then.

Before leaving, I asked Mountcastle to sign my well-read copy of *The Mindful Brain*. As I walked away, I was both happy and sad. I was happy to have met him and relieved that he thought highly of me. I felt sad knowing it was possible that I might never see him again. Even if I succeeded in my quest, I might not be able to share with him what I had learned and get his help and feedback. As I walked to my cab, I felt determined to complete his mission.

CHAPTER 3

A Model of the World in Your Head

What the brain does may seem obvious to you. The brain gets inputs from its sensors, it processes those inputs, and then it acts. In the end, how an animal reacts to what it senses determines its success or failure. A direct mapping from sensory input to action certainly applies to some parts of the brain. For example, accidentally touching a hot surface will cause a reflex retraction of the arm. The input-output circuit responsible is located in the spinal cord. But what about the neocortex? Can we say that the task of the neocortex is to take inputs from sensors and then immediately act? In short, no.

You are reading or listening to this book and it isn't causing any immediate actions other than perhaps turning pages or touching a screen. Thousands of words are streaming into your neocortex and, for the most part, you are not acting on them. Maybe later you will act differently for having read this book. Perhaps you will have future conversations about brain theory and the future of humanity that you would not have had if you didn't read this book. Perhaps your future thoughts and word choices will be subtly influenced by my words. Perhaps you will work on creating intelligent machines based on brain principles, and my words will inspire you in this direction. But right now, you are just reading. If we insist on describing the neocortex as an input-output system, then the best we could say it is that the neocortex gets lots of inputs, it learns from these inputs, and then, later—maybe hours, maybe years—it acts differently based on these prior inputs.

From the moment I became interested in how the brain worked, I realized that thinking of the neocortex as an input-leads-to-output system would not be fruitful. Fortunately, when I was a graduate student at Berkeley I had an insight that led me down a different and more successful path. I was at home, working at my desk. There were dozens of objects on the desk and in the room. I realized that if any one of these objects changed, in even the slightest way, I would notice it. My pencil cup was always on the right side of the table; if one day I found it on the left, I would notice the change and wonder how it got moved. If the stapler changed in length, I would notice. I would notice the change if I touched the stapler or if I looked at it. I would even notice if the stapler made a different sound when being used. If the clock on the wall changed its location or style, I would notice. If the cursor on my computer screen moved left when I moved the mouse to the right, I would immediately realize something was wrong. What struck me was that I would notice these changes even if I wasn't attending to these objects. As I looked around the room, I didn't ask, "Is my stapler the correct length?" I didn't think, "Check to make sure the clock's hour hand is still shorter than the minute hand." Changes to the normal would just pop into my head, and my attention would then be drawn to them. There were literally thousands of possible changes in my environment that my brain would notice almost instantly.

There was only one explanation I could think of. My brain, specifically my neocortex, was making multiple simultaneous predictions of what it was about to see, hear, and feel. Every time I moved my eyes, my neocortex made predictions of what it was about to see. Every time I picked something up, my neocortex made predictions of what each finger should feel. And every action I took led to predictions of what I should hear. My brain predicted the smallest stimuli, such as the texture of the handle on my coffee cup, and large conceptual ideas, such as the correct month that should be displayed on a calendar. These predictions occurred in every sensory modality, for low-level sensory features and high-level concepts, which told me that every part of the neocortex, and therefore every cortical column, was making predictions. Prediction was a ubiquitous function of the neocortex.

At that time, few neuroscientists described the brain as a prediction machine. Focusing on how the neocortex made many parallel predictions would be a novel way to study how it worked. I knew that prediction wasn't the only thing the neocortex did, but prediction represented a systemic way of attacking the cortical column's mysteries. I could ask specific questions about how neurons make predictions under different conditions. The answers to these questions might reveal what cortical columns do, and how they do it.

To make predictions, the brain has to learn what is normal—that is, what should be expected based on past experience. My previous book, *On Intelligence*, explored this idea of learning and prediction. In the book, I used the phrase "the memory prediction framework" to describe the overall idea, and I wrote about the implications of thinking about the brain this way. I argued that by studying how the neocortex makes predictions, we would be able to unravel how the neocortex works.

Today I no longer use the phrase "the memory prediction framework." Instead, I describe the same idea by saying that the neocortex learns a model of the world, and it makes predictions based on its model. I prefer the word "model" because it more precisely describes the kind of information that the neocortex learns. For example, my brain has a model of my stapler. The model of the stapler includes what the stapler looks like, what it feels like, and the sounds it makes when being used. The brain's model of the world includes where objects are and how they change when we interact with them. For example, my model of the stapler includes how the top of the stapler moves relative to the bottom and how a staple comes out when the top is pressed down. These actions may seem simple, but you were not born with this knowledge. You learned it at some point in your life and now it is stored in your neocortex.

The brain creates a predictive model. This just means that the brain continuously predicts what its inputs will be. Prediction isn't something that the brain does every now and then; it is an intrinsic property that never stops, and it serves an essential role in learning. When the brain's predictions are verified, that means the brain's model of the world is accurate. A mis-prediction causes you to attend to the error and update the model.

We are not aware of the vast majority of these predictions unless the input to the brain does not match. As I casually reach out to grab my coffee cup, I am not aware that my brain is predicting what each finger will feel, how heavy the cup should be, the temperature of the cup, and the sound the cup will make when I place it back on my desk. But if the cup was suddenly heavier, or cold, or squeaked, I would notice the change. We can be certain that these predictions are occurring because even a small change in any of these inputs will be noticed. But when a prediction is correct, as most will be, we won't be aware that it ever occurred.

When you are born, your neocortex knows almost nothing. It doesn't know any words, what buildings are like, how to use a computer, or what a door is and how it moves on hinges. It has to learn countless things. The overall structure of the neocortex is not random. Its size, the number of regions it has, and how they are connected together is largely determined by our genes. For example, genes determine what parts of the neocortex are connected to the eyes, what other parts are connected to the ears, and how those parts connect to each other. Therefore, we can say that the neocortex is structured at birth to see, hear, and even learn language. But it is also true that the neocortex doesn't know what it will see, what it will hear, and what specific languages it might learn. We can think of the neocortex as starting life having some built-in assumptions about the world but knowing nothing in particular. Through experience, it learns a rich and complicated model of the world.

The number of things the neocortex learns is huge. I am sitting in a room with hundreds of objects. I will randomly pick one: a printer. I have learned a model of the printer that includes it having a paper tray, and how the tray moves in and out of the printer. I know how to change the size of the paper and how to unwrap a new ream and place it in the tray. I know the steps I need to take to clear a paper jam. I know that the power cord has a D-shaped plug at one end and that it can only be inserted in one orientation. I know the sound of the printer and how that sound is different when it is printing on two sides of a sheet of paper rather than on one. Another object in my room is a small, two-drawer file cabinet. I can recall dozens of things I know about the cabinet, including what is in each drawer and how the objects in the drawer are arranged. I know there is a lock, where the key is, and how to insert and turn the key to lock the cabinet. I know how the key and lock feel and the sounds they make as I use them. The key has a small ring attached to it and I know how to use my fingernail to pry open the ring to add or remove keys.

Imagine going room to room in your home. In each room you can think of hundreds of things, and for each item you can follow a cascade of learned knowledge. You can also do the same exercise for the town you live in, recalling what buildings, parks, bike racks, and individual trees exist at different locations. For each item, you can recall experiences associated with it and how you interact with it. The number of things you know is enormous, and the associated links of knowledge seem never-ending.

We learn many high-level concepts too. It is estimated that each of us knows about forty thousand words. We have the ability to learn spoken language, written language, sign language, the language of mathematics, and the language of music. We learn how

electronic forms work, what thermostats do, and even what empa-
thy or democracy mean, although our understanding of these may
differ. Independent of what other things the neocortex might do,
we can say for certain that it learns an incredibly complex model
of the world. This model is the basis of our predictions, perceptions, and actions.

Learning Through Movement

The inputs to the brain are constantly changing. There are two reasons why. First, the world can change. For example, when listening to music, the inputs from the ears change rapidly, reflecting
the movement of the music. Similarly, a tree swaying in the breeze
will lead to visual and perhaps auditory changes. In these two examples, the inputs to the brain are changing from moment to moment, not because you are moving but because things in the world
are moving and changing on their own.

The second reason is because we move. Every time we take a
step, move a limb, move our eyes, tilt our head, or utter a sound,
the input from our sensors change. For example, our eyes make
rapid movements, called saccades, about three times a second.
With each saccade, our eyes fixate on a new point in the world and
the information from the eyes to the brain changes completely.
This change would not occur if we hadn't moved our eyes.

The brain learns its model of the world by observing how its
inputs change over time. There isn't another way to learn. Unlike
with a computer, we cannot upload a file into our brain. The only
way for a brain to learn anything is via changes in its inputs. If the
inputs to the brain were static, nothing could be learned.

Some things, like a melody, can be learned without moving the
body. We can sit perfectly still, with eyes closed, and learn a new
melody by just listening to how the sounds change over time. But
most learning requires that we actively move and explore. Imagine you enter a new house, one you have not been in before. If
you don't move, there will be no changes in your sensory input,

and you can't possibly learn anything about the house. To learn a model of the house, you have to look in different directions and walk from room to room. You need to open doors, peek in drawers, and pick up objects. The house and its contents are mostly static; they don't move on their own. To learn a model of a house, you have to move.

Take a simple object such as a computer mouse. To learn what a mouse feels like, you have to run your fingers over it. To learn what a mouse looks like, you have to look at it from different angles and fixate your eyes on different locations. To learn what a mouse does, you have to press down on its buttons, slide off the battery cover, or move it across a mouse pad to see, feel, and hear what happens.

The term for this is sensory-motor learning. In other words, the brain learns a model of the world by observing how our sensory inputs change as we move. We can learn a song without moving because, unlike the order in which we can move from room to room in a house, the order of notes in a song is fixed. But most of the world isn't like that; most of the time we have to move to discover the structure of objects, places, and actions. With sensory-motor learning, unlike a melody, the order of sensations is not fixed. What I see when I enter a room depends on which direction I turn my head. What my finger feels when holding a coffee cup depends on whether I move my finger up or down or sideways.

With each movement, the neocortex predicts what the next sensation will be. Move my finger up on the coffee cup and I expect to feel the lip, move my finger sideways and I expect to feel the handle. If I turn my head left when entering my kitchen, I expect to see my refrigerator, and if I turn my head right, I expect to see the range. If I move my eyes to the left front burner, I expect to see the broken igniter that I need to fix. If any input doesn't match the brain's prediction—perhaps my spouse fixed the igniter—then my attention is drawn to the area of mis-prediction. This alerts the neocortex that its model of that part of the world needs to be updated.

The question of how the neocortex works can now be phrased more precisely: *How does the neocortex, which is composed of thousands of nearly identical cortical columns, learn a predictive model of the world through movement?*

This is the question my team and I set out to answer. Our belief was that if we could answer it, we could reverse engineer the neocortex. We would understand both what the neocortex did and how it did it. And ultimately, we would be able to build machines that worked the same way.

Two Tenets of Neuroscience

Before we can start answering the question above, there are a few more basic ideas you need to know. First, like every other part of the body, the brain is composed of cells. The brain's cells, called neurons, are in many ways similar to all our other cells. For example, a neuron has a cell membrane that defines its boundary and a nucleus that contains DNA. However, neurons have several unique properties that don't exist in other cells in your body.

The first is that neurons look like trees. They have branch-like extensions of the cell membrane, called axons and dendrites. The dendrite branches are clustered near the cell and collect the inputs. The axon is the output. It makes many connections to nearby neurons but often travels long distances, such as from one side of the brain to the other or from the neocortex all the way down to the spinal cord.

The second difference is that neurons create spikes, also called action potentials. An action potential is an electrical signal that starts near the cell body and travels along the axon until it reaches the end of every branch.

The third unique property is that the axon of one neuron makes connections to the dendrites of other neurons. The connection points are called synapses. When a spike traveling along an axon reaches a synapse, it releases a chemical that enters the dendrite of the receiving neuron. Depending on which chemical is released, it makes the receiving neuron more or less likely to generate its own spike.

Considering how neurons work, we can state two fundamental tenets. These tenets will play important roles in our understanding of the brain and intelligence.

Tenet Number One: Thoughts, Ideas, and Perceptions Are the Activity of Neurons

At any point in time, some neurons in the neocortex are actively spiking and some are not. Typically, the number of neurons that are active at the same time is small, maybe 2 percent. Your thoughts and perceptions are determined by which neurons are spiking. For example, when doctors perform brain surgery, they sometimes need to activate neurons in an awake patient's brain. They stick a tiny probe into the neocortex and use electricity to activate a few neurons. When they do this, the patient might hear, see, or think something. When the doctor stops the stimulation, whatever the patient was experiencing stops. If the doctor activates different neurons, the patient has a different thought or perception.

Thoughts and experiences are always the result of a set of neurons that are active at the same time. Individual neurons can participate in many different thoughts or experiences. Every thought you have is the activity of neurons. Everything you see, hear, or feel is also the activity of neurons. Our mental states and the activity of neurons are one and the same.

Tenet Number Two: Everything We Know Is Stored in the Connections Between Neurons

The brain remembers a lot of things. You have permanent memories, such as where you grew up. You have temporary memories, such as what you had for dinner last night. And you have basic knowledge, such as how to open a door or how to spell the word "dictionary." All these things are stored using synapses, the connections between neurons.

Here is the basic idea for how the brain learns: Each neuron has thousands of synapses, which connect the neuron to thousands of other neurons. If two neurons spike at the same time, they will strengthen the connection between them. When we learn something, the connections are strengthened, and when we forget something, the connections are weakened. This basic idea was proposed by Donald Hebb in the 1940s and today it is referred to as Hebbian learning.

For many years, it was believed that the connections between neurons in an adult brain were fixed. Learning, it was believed, involved increasing or decreasing the strength of synapses. This is still how learning occurs in most artificial neural networks.

However, over the past few decades, scientists have discovered that in many parts of the brain, including the neocortex, new synapses form and old ones disappear. Every day, many of the synapses on an individual neuron will disappear and new ones will replace them. Thus, much of learning occurs by forming new connections between neurons that were previously not connected. Forgetting happens when old or unused connections are removed entirely.

The connections in our brain store the model of the world that we have learned through our experiences. Every day we experience new things and add new pieces of knowledge to the model by forming new synapses. The neurons that are active at any point in time represent our current thoughts and perceptions.

We have now gone over several of the basic building blocks of the neocortex—some of the pieces of our puzzle. In the next chapter, we start putting these pieces together to reveal how the entire neocortex works.

CHAPTER 4

The Brain Reveals Its Secrets

People often say the brain is the most complicated thing in the universe. They conclude from this that there will not be a simple explanation for how it works, or that perhaps we will never understand it. The history of scientific discovery suggests they are wrong. Major discoveries are almost always preceded by bewildering, complex observations. With the correct theoretical framework, the complexity does not disappear, but it no longer seems confusing or daunting.

A familiar example is the movement of the planets. For thousands of years, astronomers carefully tracked the motion of the planets among the stars. The path of a planet over the course of a year is complex, darting this way and that, making loops in the sky. It was hard to imagine an explanation for these wild movements. Today, every child learns the basic idea that the planets orbit the Sun. The motion of the planets is still complex, and to predict their course requires difficult mathematics, but with the right framework, the complexity is no longer mysterious. Few scientific discoveries are hard to understand at a basic level. A child can learn that the Earth orbits the Sun. A high school student can learn the principles of evolution, genetics, quantum mechanics, and relativity. Each of these scientific advances was preceded by confusing observations. But now, they seem straightforward and logical.

Similarly, I always believed that the neocortex appeared complicated largely because we didn't understand it, and that it would appear relatively simple in hindsight. Once we knew the solution, we would look back and say, "Oh, of course, why didn't we think of that?" When our research stalled or when I was told that the brain is too difficult to understand, I would imagine a future where brain theory was part of every high school curriculum. This kept me motivated.

Our progress in trying to decipher the neocortex had its ups and downs. Over the course of eighteen years—three at the Redwood Neuroscience Institute and fifteen at Numenta—my colleagues and I worked on this problem. There were times when we made small advances, times we made large advances, and times we pursued ideas that at first seemed exciting but ultimately proved to be dead ends. I am not going to walk you through all of this history. Instead, I want to describe several key moments when our understanding took a leap forward, when nature whispered in our ear telling us something we had overlooked. There are three such "aha" moments that I remember vividly.

Discovery Number One: The Neocortex Learns a Predictive Model of the World

I already described how, in 1986, I realized that the neocortex learns a predictive model of the world. I can't overstate the importance of this idea. I call it a discovery because that is how it felt to me at the time. There is a long history of philosophers and scientists talking about related ideas, and today it is not uncommon for neuroscientists to say the brain learns a predictive model of the world. But in 1986, neuroscientists and textbooks still described the brain more like a computer; information comes in, it gets processed, and then the brain acts. Of course, learning a model of the world and making predictions isn't the only thing the neocortex does. However, by studying how the neocortex makes predictions, I believed we could unravel how the entire system worked.

This discovery led to an important question. How does the brain make predictions? One potential answer is that the brain has two types of neurons: neurons that fire when the brain is actually seeing something, and neurons that fire when the brain is predicting it will see something. To avoid hallucinating, the brain needs to keep its predictions separate from reality. Using two sets of neurons does this nicely. However, there are two problems with this idea.

First, given that the neocortex is making a massive number of predictions at every moment, we would expect to find a large number of prediction neurons. So far, that hasn't been observed. Scientists have found some neurons that become active in advance of an input, but these neurons are not as common as we would expect. The second problem is based on an observation that had long bothered me. If the neocortex is making hundreds or thousands of predictions at any moment in time, why are we not aware of most of these predictions? If I grab a cup with my hand, I am not aware that my brain is predicting what each finger should feel, unless I feel something unusual—say, a crack. We are not consciously aware of most of the predictions made by the brain unless an error occurs. Trying to understand how the neurons in the neocortex make predictions led to the second discovery.

Discovery Number Two: Predictions Occur Inside Neurons

Recall that predictions made by the neocortex come in two forms. One type occurs because the world is changing around you. For example, you are listening to a melody. You can be sitting still with your eyes closed and the sound entering your ears changes as the melody progresses. If you know the melody, then your brain continually predicts the next note and you will notice if any of the notes are incorrect. The second type of prediction occurs because you are moving relative to the world. For example, as I lock my bicycle in the lobby of my office, my neocortex makes many predictions about what I will feel, see, and hear based on my movements. The bicycle and lock don't move on their own. Every action I make leads to a set of predictions. If I change the order of my actions, the order of the predictions also changes.

Mountcastle's proposal of a common cortical algorithm suggested that every column in the neocortex makes both types of predictions. Otherwise, cortical columns would have differing functions. My team also realized that the two types of prediction are closely related. Therefore, we felt that progress on one subproblem would lead to progress on the other.

Predicting the next note in a melody, also known as sequence memory, is the simpler of the two problems, so we worked on it first. Sequence memory is used for a lot more than just learning melodies; it is also used in creating behaviors. For example, when I dry myself off with a towel after showering, I typically follow a nearly identical pattern of movements, which is a form of sequence memory. Sequence memory is also used in language. Recognizing a spoken word is like recognizing a short melody. The word is defined by a sequence of phonemes, whereas a melody is defined by a sequence of musical intervals. There are many more examples, but for simplicity I will stick to melodies. By deducing how neurons in a cortical column learn sequences, we hoped to discover basic principles of how neurons make predictions about everything.

We worked on the melody-prediction problem for several years before we were able to deduce the solution, which had to exhibit numerous capabilities. For example, melodies often have repeating sections, such as a chorus or the *da da da dum* of Beethoven's Fifth Symphony. To predict the next note, you can't just look at the previous note or the previous five notes. The correct prediction may rely on notes that occurred a long time ago. Neurons have to figure out how much context is necessary to make the right prediction. Another requirement is that neurons have to play Name That Tune. The first few notes you hear might belong to several different melodies. The neurons have to keep track of all possible melodies consistent with what has been heard so far, until enough notes have been heard to eliminate all but one melody.

Engineering a solution to the sequence-memory problem would be easy, but figuring out how real neurons—arranged as we see them in the neocortex—solve these and other requirements was hard. Over several years, we tried different approaches. Most worked to some extent, but none exhibited all the capabilities we needed and none precisely fit the biological details we knew about the brain. We were not interested in a partial solution or in a "biologically inspired" solution. We wanted to know exactly how real neurons, arranged as seen in the neocortex, learn sequences and make predictions.

I remember the moment when I came upon the solution to the melody-prediction problem. It was in 2010, one day prior to our Thanksgiving holiday. The solution came in a flash. But as I thought about it, I realized it required neurons to do things I wasn't sure they were capable of. In other words, my hypothesis made several detailed and surprising predictions that I could test.

Scientists normally test a theory by running experiments to see if the predictions made by the theory hold up or not. But neuroscience is unusual. There are hundreds to thousands of published papers in every subfield, and most of these papers present experimental data that are unassimilated into any overall theory. This provides theorists like myself an opportunity to quickly test a new hypothesis by searching through past research to find experimental evidence that supports or invalidates it. I found a few dozen journal papers that contained experimental data that could shed light on the new sequence-memory theory. My extended family was staying over for the holiday, but I was too excited to wait until everyone went home. I recall reading papers while cooking and engaging my relatives in discussions about neurons and melodies. The more I read, the more confident I became that I had discovered something important.

The key insight was a new way of thinking about neurons.

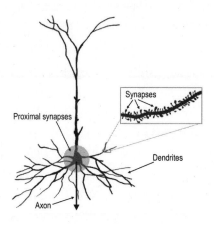

Proximal synapses

Synapses

Dendrites

Axon

A typical neuron

Above is a picture of the most common type of neuron in the neocortex. Neurons like this have thousands, sometimes tens of thousands, of synapses spaced along the branches of the dendrites. Some of the dendrites are near the cell body (which is toward the bottom of the image), and some dendrites are farther away (toward the top). The box shows an enlarged view of one dendrite branch so you can see how small and tightly packed the synapses are. Each bump along the dendrite is one synapse. I've also highlighted an area around the cell body; the synapses in this area are called proximal synapses. If the proximal synapses receive enough input, then the neuron will spike. The spike starts at the cell body and travels to other neurons via the axon. The axon was not visible in this picture, so I added a down-facing arrow to show where it would be. If you just consider the proximal synapses and the cell body, then this is the classic view of a neuron. If you have ever read about neurons or studied artificial neural networks, you will recognize this description.

Oddly, less than 10 percent of the cell's synapses are in the proximal area. The other 90 percent are too far away to cause a spike. If an input arrives at one of these distal synapses, like the ones shown in the box, it has almost no effect on the cell body. All that researchers could say was that the distal synapses performed some sort of modulatory role. For many years, no one knew what 90 percent of the synapses in the neocortex did.

Starting around 1990, this picture changed. Scientists discovered new types of spikes that travel along the dendrites. Before, we knew of only one type of spike: it started at the cell body and traveled along the axon to reach other cells. Now, we'd learned that there were other spikes that traveled along the dendrites. One type of dendrite spike begins when a group of twenty or so synapses situated next to each other on a dendrite branch receive input at the same time. Once a dendrite spike is activated, it travels along the dendrite until it reaches the cell body. When it gets there, it raises the voltage of the cell, but not enough to make the neuron spike. It is like the dendrite spike is teasing the neuron—it is almost strong enough to make the neuron active, but not quite.

The neuron stays in this provoked state for a little bit of time before going back to normal. Scientists were once again puzzled. What are dendrite spikes good for if they aren't powerful enough to create a spike at the cell body? Not knowing what dendrite spikes are for, AI researchers use simulated neurons that don't have them. They also don't have dendrites and the many thousands of synapses found on the dendrites. I knew that distal synapses had to play an essential role in brain function. Any theory and any neural network that did not account for 90 percent of the synapses in the brain had to be wrong.

The big insight I had was that dendrite spikes are predictions. A dendrite spike occurs when a set of synapses close to each other on a distal dendrite get input at the same time, and it means that the neuron has recognized a pattern of activity in some other neurons. When the pattern of activity is detected, it creates a dendrite spike, which raises the voltage at the cell body, putting the cell into what we call a predictive state. The neuron is then primed to spike. It is similar to how a runner who hears "Ready, set . . ." is primed to start running. If a neuron in a predictive state subsequently gets enough proximal input to create an action potential spike, then the cell spikes a little bit sooner than it would have if the neuron was not in a predictive state.

Imagine there are ten neurons that all recognize the same pattern on their proximal synapses. This is like ten runners on a starting line, all waiting for the same signal to begin racing. One runner hears "Ready, set . . ." and anticipates the race is about to begin. She gets in the blocks and is primed to start. When the "go" signal is heard, she gets off the blocks sooner than the other runners who weren't primed, who didn't hear a preparatory signal. Upon seeing the first runner off to an early lead, the other runners give up and don't even start. They wait for the next race. This kind of competition occurs throughout the neocortex.

In each minicolumn, multiple neurons respond to the same input pattern. They are like the runners on the starting line, all waiting for the same signal. If their preferred input arrives, they all want to start spiking. However, if one or more of the neurons are in the predictive state, our theory says, only those neurons spike and the other neurons are inhibited. Thus, when an input arrives that is unexpected, multiple neurons fire at once. If the input is predicted, then only the predictive-state neurons become active. This is a common observation about the neocortex: unexpected inputs cause a lot more activity than expected ones.

If you take several thousand neurons, arrange them in minicolumns, let them make connections to each other, and add a few inhibitory neurons, they learn sequences. The neurons solve the Name That Tune problem, they don't get confused by repeated subsequences, and, collectively, they predict the next element in the sequence.

The trick to making this work was a new understanding of the neuron. We previously knew that prediction is a ubiquitous function of the brain. But we didn't know how or where predictions are made. With this discovery, we understood that most predictions occur inside neurons. A prediction occurs when a neuron recognizes a pattern, creates a dendrite spike, and is primed to spike earlier than other neurons. With thousands of distal synapses, each neuron can recognize hundreds of patterns that predict when the neuron should become active. Prediction is built into the fabric of the neocortex, the neuron.

We spent over a year testing the new neuron model and sequence-memory circuit. We wrote software simulations that tested its capacity and were surprised to find that as few as twenty thousand neurons can learn thousands of complete sequences. We found that the sequence memory continued to work even if 30 percent of the neurons died or if the input was noisy. The more time we spent testing our theory, the more confidence we gained that it truly captured what was happening in the neocortex. We also found increasing empirical evidence from experimental labs that supported our idea. For example, the theory predicts that dendrite spikes behave in some specific ways, but at first, we couldn't find conclusive experimental evidence. However, by talking to experimentalists we were able to get a clearer understanding of their findings and see that the data were consistent with what we predicted. We first published the theory in a white paper in 2011. We followed this with a peer-reviewed journal paper in 2016, titled "Why Neurons Have Thousands of Synapses, a Theory of Sequence Memory in the Neocortex." The reaction to the paper was heartening, as it quickly became the most read paper in its journal.

Discovery Number Three: The Secret of the Cortical Column Is Reference Frames

Next, we turned our attention to the second half of the prediction problem: How does the neocortex predict the next input when we move? Unlike a melody, the order of inputs in this situation is not fixed, as it depends on which way we move. For example, if I look left, I see one thing; if I look right, I see something else. For a corti-cal column to predict its next input, it must know what movement is about to occur.

Predicting the next input in a sequence and predicting the next input when we move are similar problems. We realized that our sequence-memory circuit could make both types of predictions if the neurons were given an additional input that represented how the sensor was moving. However, we did not know what the movement-related signal should look like.

We started with the simplest thing we could think of: What if the movement-related signal was just "move left" or "move right"? We tested this idea, and it worked. We even built a little robot arm that predicted its input as it moved left and right, and we demon-strated it at a neuroscience conference. However, our robot arm had limitations. It worked for simple problems, such as moving in two directions, but when we tried to scale it to work with the complexity of the real world, such as moving in multiple direc-tions at the same time, it required too much training. We felt we were close to the correct solution, but something was wrong. We tried several variations with no success. It was frustrating. After several months, we still could not see a way to solve the problem, so we put this question aside and worked on other things for a while.

Toward the end of February 2016, I was in my office waiting for my spouse, Janet, to join me for lunch. I was holding a Nu-menta coffee cup in my hand and observed my fingers touching it. I asked myself a simple question: What does my brain need to know to predict what my fingers will feel as they move? If one of my fingers is on the side of the cup and I move it toward the top, my brain predicts that I will feel the rounded curve of the lip. My brain makes this prediction before my finger touches the lip. What does the brain need to know to make this prediction? The answer was easy to state. The brain needs to know two things: *what* object it is touching (in this case the coffee cup) and *where* my finger will be on the cup after my finger moves.

Notice that the brain needs to know where my finger is relative to the cup. It doesn't matter where my finger is relative to my body, and it doesn't matter where the cup is or how it is positioned. The cup can be tilted left or tilted right. It could be in front of me or off to the side. What matters is the location of my finger *relative to the cup.*

This observation means there must be neurons in the neocortex that represent the location of my finger in a reference frame that is attached to the cup. The movement-related signal we had been searching for, the signal we needed to predict the next input, was "location on the object."

You probably learned about reference frames in high school. The x, y, and z axes that define the location of something in space are an example of a reference frame. Another familiar example is latitude and longitude, which define locations on the surface of Earth. At first, it was hard for us to imagine how neurons could represent something like x, y, and z coordinates. But even more puzzling was that neurons could attach a reference frame to an object like a coffee cup. The cup's reference frame is relative to the cup; therefore, the reference frame must move with the cup.

Imagine an office chair. My brain predicts what I will feel when I touch the chair, just as my brain predicts what I will feel when I touch the coffee cup. Therefore, there must be neurons in my neocortex that know the location of my finger relative to the chair, meaning that my neocortex must establish a reference frame that is fixed to the chair. If I spin the chair in a circle, the reference frame spins with it. If I flip the chair over, the reference frame flips over. You can think of the reference frame as an invisible three-dimensional grid surrounding, and attached to, the chair. Neurons are simple things. It was hard to imagine that they could create and attach reference frames to objects, even as those objects were moving and rotating out in the world. But it got even more surprising.

Different parts of my body (fingertips, palm, lips) might touch the coffee cup at the same time. Each part of my body that touches the cup makes a separate prediction of what it will feel based on its unique location on the cup. Therefore, the brain isn't making one prediction; it's making dozens or even hundreds of predictions at the same time. The neocortex must know the location, relative to the cup, of every part of my body that is touching it.

Vision, I realized, is doing the same thing as touch. Patches of retina are analogous to patches of skin. Each patch of your retina *sees* only a small part of an entire object, in the same way that each patch of your skin *touches* only a small part of an object. The brain doesn't process a picture; it starts with a picture on the back of the eye but then breaks it up into hundreds of pieces. It then assigns each piece to a location relative to the object being observed.

Creating reference frames and tracking locations is not a trivial task. I knew it would take several different types of neurons and multiple layers of cells to make these calculations. Since the complex circuitry in every cortical column is similar, locations and reference frames must be universal properties of the neocortex. Each column in the neocortex—whether it represents visual input, tactile input, auditory input, language, or high-level thought—must have neurons that represent reference frames and locations.

Up to that point, most neuroscientists, including me, thought that the neocortex primarily processed sensory input. What I realized that day is that we need to think of the neocortex as primarily processing reference frames. Most of the circuitry is there to create reference frames and track locations. Sensory input is of course essential. As I will explain in coming chapters, the brain builds models of the world by associating sensory input with locations in reference frames.

Why are reference frames so important? What does the brain gain from having them? First, a reference frame allows the brain to learn the structure of something. A coffee cup is a thing because it is composed of a set of features and surfaces arranged relative to each other in space. Similarly, a face is a nose, eyes, and mouth arranged in relative positions. You need a reference frame to specify the relative positions and structure of objects.

Second, by defining an object using a reference frame, the brain can manipulate the entire object at once. For example, a car has many features arranged relative to each other. Once we learn a car, we can imagine what it looks like from different points of view or if it were stretched in one dimension. To accomplish these feats, the brain only has to rotate or stretch the reference frame and all the features of the car rotate and stretch with it.

Third, a reference frame is needed to plan and create movements. Say my finger is touching the front of my phone and I want to press the power button at the top. If my brain knows the current location of my finger and the location of the power button, then it can calculate the movement needed to get my finger from its current location to the desired new one. A reference frame relative to the phone is needed to make this calculation.

Reference frames are used in many fields. Roboticists rely on them to plan the movements of a robot's arm or body. Reference frames are also used in animated films to render characters as they move. A few people had suggested that reference frames might be needed for certain AI applications. But as far as I know, there had not been any significant discussion that the neocortex worked on reference frames, and that the function of most of the neurons in each cortical column is to create reference frames and track locations. Now it seems obvious to me.

Vernon Mountcastle argued there was a universal algorithm that exists in every cortical column, yet he didn't know what the algorithm was. Francis Crick wrote that we needed a new framework to understand the brain, yet he, too, didn't know what that framework should be. That day in 2016, holding the cup in my hand, I realized Mountcastle's algorithm and Crick's framework were both based on reference frames. I didn't yet understand how neurons could do this, but I knew it must be true. Reference frames were the missing ingredient, the key to unraveling the mystery of the neocortex and to understanding intelligence.

All these ideas about locations and reference frames occurred to me in what seemed like a second. I was so excited that I jumped out of my chair and ran to tell my colleague Subutai Ahmad. As I raced the twenty feet to his desk, I ran into Janet and almost knocked her over. I was anxious to talk to Subutai, but while I was steadying Janet and apologizing to her, I realized it would be wiser to talk to him later. Janet and I discussed reference frames and locations as we shared a frozen yogurt.

> This is a good point to address a question I am often asked: How can I speak confidently about a theory if it hasn't been tested experimentally? I just described one of these situations. I had an insight that the neocortex is infused with reference frames, and I immediately started talking about it with certainty. As I write this book, there is growing evidence to support this new idea, but it still has not been thoroughly tested. And yet, I have no hesitation describing this idea as a fact. Here is why.
>
> As we work on a problem, we uncover what I call constraints. Constraints are things that the solution to the problem must address. I gave a few examples of constraints when describing sequence memory, for example, the Name That Tune requirement. The anatomy and physiology of the brain are also constraints. Brain theory must ultimately explain all the details of the brain, and a correct theory cannot violate any of those details.
>
> The longer you work on a problem, the more constraints you discover and the harder it becomes to imagine a solution. The aha moments I described in this chapter were about problems that we worked on for years. Therefore, we understood these problems deeply and our list of constraints was long. The likelihood that a solution is correct increases exponentially with the number of constraints it satisfies. It is like solving a crossword puzzle: There are often several words that match an individual clue. If you pick one of those words, it could be wrong. If you find two intersecting words that work, then it is much more likely they are both correct.

If you find ten intersecting words, the chance that they are all wrong is miniscule. You can write the answer in ink without any worries.

Aha moments occur when a new idea satisfies multiple constraints. The longer you have worked on a problem—and, consequently, the more constraints the solution resolves—the bigger the aha feeling and the more confident you are in trusting the answer. The idea that the neocortex is infused with reference frames solved so many constraints that I immediately knew it was correct.

It took us over three years to work out the implications of this discovery, and, as I write, we still are not done. We have published several papers on it so far. The first paper is titled "A Theory of How Columns in the Neocortex Enable Learning the Structure of the World." This paper starts with the same circuit we described in the 2016 paper on neurons and sequence memory. We then added one layer of neurons representing location and a second layer representing the object being sensed. With these additions, we showed that a *single* cortical column could learn the three-dimensional shape of objects by sensing and moving and sensing and moving.

For example, imagine reaching into a black box and touching a novel object with one finger. You can learn the shape of the entire object by moving your finger over its edges. Our paper explained how a single cortical column can do this. We also showed how a column can recognize a previously learned object in the same manner, for example by moving a finger. We then showed how multiple columns in the neocortex work together to more quickly recognize objects. For example, if you reach into the black box and grab an unknown object with your entire hand, you can recognize it with fewer movements and in some cases in a single grasp.

We were nervous about submitting this paper and debated whether we should wait. We were proposing that the entire neocortex worked by creating reference frames, with many thousands active simultaneously. This was a radical idea. And yet, we had no proposal for how neurons actually created reference frames. Our argument was something like, "We deduced that locations and reference frames must exist and, assuming they do exist, here is how a cortical column might work. And, oh, by the way, we don't know how neurons could actually create reference frames." We decided to submit the paper anyway. I asked myself, Would I want to read this paper even though it was incomplete? My answer was yes. The idea that the neocortex represents locations and reference frames in every column was too exciting to hold back just because we didn't know how neurons did it. I was confident the basic idea was correct.

It takes a long time to put together a paper. The prose alone can take months to write, and there are often simulations to run, which can take additional months. Toward the end of this process, I had an idea that we added to the paper just prior to submitting. I suggested that we might find the answer to how neurons in the neocortex create reference frames by looking at an older part of the brain called the entorhinal cortex. By the time the paper was accepted a few months later, we knew this conjecture was correct, as I will discuss in the next chapter.

We just covered a lot of ground, so let's do a quick review. The goal of this chapter was to introduce you to the idea that every cortical column in the neocortex creates reference frames. I walked you through the steps we took to reach this conclusion. We started with the idea that the neocortex learns a rich and detailed model of the world, which it uses to constantly predict what its next sensory inputs will be. We then asked how neurons can make these predictions. This led us to a new theory that most predictions are represented by dendrite spikes that temporarily change the voltage inside a neuron and make a neuron fire a little bit sooner than it would otherwise. Predictions are not sent along the cell's axon to other neurons, which explains why we are unaware of most of them. We then showed how circuits in the neocortex that use the new neuron model can learn and predict sequences. We applied this idea to the question of how such a circuit could predict the next sensory input when the inputs are changing due to our own movements. To make these sensory-motor predictions, we deduced that each cortical column must know the location of its input relative to the object being sensed. To do that, a cortical column requires a reference frame that is fixed to the object.

CHAPTER 5

Maps in the Brain

It took years for us to deduce that reference frames exist throughout the neocortex, but in hindsight, we could have understood this a long time ago with a simple observation. Right now, I am sitting in a small lounge area of Numenta's office. Near me are three comfortable chairs similar to the one I am sitting in. Beyond the chairs are several freestanding desks. Beyond the desks, I see the old county courthouse across the street. Light from these objects enters my eyes and is projected onto the retina. Cells in the retina convert light into spikes. This is where vision starts, at the back of the eye. Why, then, do we not perceive objects as being in the eye? If the chairs, desks, and courthouse are imaged next to each other on my retina, how is it that I perceive them to be at different distances and different locations? Similarly, if I hear a car approaching, why do I perceive the car as one hundred feet away to my right and not in my ear, where the sound actually is?

This simple observation, that we perceive objects as being somewhere—not in our eyes and ears, but at some location out in the world—tells us that the brain must have neurons whose activity represents the location of every object that we perceive.

At the end of the last chapter, I told you that we were worried about submitting our first paper about reference frames because, at that time, we didn't know how neurons in the neocortex could do this. We were proposing a major new theory about how the neocortex works, but the theory was largely based on logical deduction. It would be a stronger paper if we could show how neurons did it. The day before we submitted, I added a few lines of text suggesting that the answer might be found in an older part of the brain called the entorhinal cortex. I am going to tell you why we suggested that with a story about evolution.

An Evolutionary Tale

When animals first started moving about in the world, they needed a mechanism to decide which way to move. Simple animals have simple mechanisms. For example, some bacteria follow gradients. If the quantity of a needed resource, such as food, is increasing, then they are more likely to keep moving in the same direction. If the quantity is decreasing, then they are more likely to turn and try a different direction. A bacterium doesn't know where it is; it doesn't have any way to represent its location in the world. It just goes forward and uses a simple rule for deciding when to turn. A slightly more sophisticated animal, such as an earthworm, might move to stay within desirable ranges of warmth, food, and water, but it doesn't know where it is in the garden. It doesn't know how far away the brick path is, or the direction and distance to the nearest fence post.

Now consider the advantages afforded to an animal that knows where it is, an animal that always knows its location relative to its environment. The animal can remember where it found food in the past and the places it used for shelter. The animal can then calculate how to get from its current location to these and other previously visited locations. The animal can remember the path it traveled to the watering hole and what happened at various locations along the way. Knowing your location and the location of other things in the world has many advantages, but it requires a reference frame.

Recall that a reference frame is like the grid of a map. For example, on a paper map you might locate something using labeled rows and columns, such as row D and column 7. The rows and columns of a map are a reference frame for the area represented by the map. If an animal has a reference frame for its world, then as it explores it can note what it found at each location. When the animal wants to get someplace, such as a shelter, it can use the reference frame to figure out how to get there from its current location. Having a reference frame for your world is useful for survival.

Being able to navigate the world is so valuable that evolution discovered multiple methods for doing it. For example, some honeybees can communicate distance and direction using a form of dance. Mammals, such as ourselves, have a powerful internal navigation system. There are neurons in the old part of our brain that are known to learn maps of the places we have visited, and these neurons have been under evolutionary pressure for so long that they are fine-tuned to do what they do. In mammals, the old brain parts where these map-creating neurons exist are called the hippocampus and the entorhinal cortex. In humans, these organs are roughly the size of a finger. There is one set on each side of the brain, near the center.

Maps in the Old Brain

In 1971, scientist John O'Keefe and his student Jonathan Dostrovsky placed a wire into a rat's brain. The wire recorded the spiking activity of a single neuron in the hippocampus. The wire went up toward the ceiling so they could record the activity of the cell as the rat moved and explored its environment, which was typically a big box on a table. They discovered what are now called place cells: neurons that fire every time the rat is in a particular location in a particular environment. A place cell is like a "you are here" marker on a map. As the rat moves, different place cells become active in each new location. If the rat returns to a location where it was before, the same place cell becomes active again.

In 2005, scientists in the lab of May-Britt Moser and Edvard Moser used a similar experimental setup, again with rats. In their experiments, they recorded signals from neurons in the entorhinal cortex, adjacent to the hippocampus. They discovered what are now called grid cells, which fire at multiple locations in an environment. The locations where a grid cell becomes active form a grid pattern. If the rat moves in a straight line, the same grid cell becomes active again and again, at equally spaced intervals.

The details of how place cells and grid cells work are complicated and still not completely understood, but you can think of them as creating a map of the environment occupied by the rat. Grid cells are like the rows and columns of a paper map, but overlaid on the animal's environment. They allow the animal to know where it is, to predict where it will be when it moves, and to plan movements. For example, if I am at location B4 on a map and want to get to location D6, I can use the map's grid to know that I have to go two squares to the right and two squares down.

But grid cells alone don't tell you what is at a location. For example, if I told you that you were at location A6 on a map, that information doesn't tell you what you will find there. To know what is at A6, you need to look at the map and see what is printed in the corresponding square. Place cells are like the details printed in the square. Which place cells become active depends on what the rat senses at a particular location. Place cells tell the rat where it is based on sensory input, but place cells alone aren't useful for planning movements—that requires grid cells. The two types of cells work together to create a complete model of the rat's environment.

Every time a rat enters an environment, the grid cells establish a reference frame. If it is a novel environment, the grid cells create a new reference frame. If the rat recognizes the environment, the grid cells reestablish the previously used reference frame. This process is analogous to you entering a town. If you look around and realize that you have been there before, you pull out the correct map for that town. If the town looks unfamiliar, then you take out a blank piece of paper and start creating a new map. As you walk around the town, you write on your map what you see at each location. That is what grid cells and place cells do. They create unique maps for every environment. As a rat moves, the active grid cells and the active place cells change to reflect the new location.

Humans have grid cells and place cells too. Unless you are completely disoriented, you always have a sense of where you are. I am now standing in my office. Even if I close my eyes, my sense of location persists, and I continue to know where I am. Keeping my eyes closed, I take two steps to my right and my sense of location in the room changes. The grid cells and place cells in my brain have created a map of my office, and they keep track of where I am in my office, even when my eyes are closed. As I walk, which cells are active changes to reflect my new location. Humans, rats, indeed all mammals use the same mechanism for knowing our location. We all have grid cells and place cells that create models of the places we have been.

Maps in the New Brain

When we were writing our 2017 paper about locations and reference frames in the neocortex, I had some knowledge of place cells and grid cells. It occurred to me that knowing the location of my finger relative to a coffee cup is similar to knowing the location of my body relative to a room. My finger moves around the cup in the same way that my body moves about a room. I realized that the neocortex might have neurons that are equivalent to the ones in the hippocampus and entorhinal cortex. These cortical place cells and cortical grid cells would learn models of objects in a similar way to how place cells and grid cells in the old brain learn models of environments.

Given their role in basic navigation, place cells and grid cells are almost certainly evolutionarily older than the neocortex. Therefore, I figured it was more likely that the neocortex creates reference frames using a derivative of grid cells than that it evolved a new mechanism from scratch. But in 2017, we were not aware of any evidence that the neocortex had anything similar to grid cells or place cells—it was informed speculation.

Shortly after our 2017 paper was accepted, we learned of recent experiments that suggested grid cells might be present in parts of the neocortex. (I will discuss these experiments in Chapter 7.) This was encouraging. The more we studied the literature related to grid cells and place cells, the more confident we became that cells that perform similar functions exist in every cortical column. We first made this argument in a 2019 paper, titled "A Framework for Intelligence and Cortical Function Based on Grid Cells in the Neocortex."

Again, to learn a complete model of something you need both grid cells and place cells. Grid cells create a reference frame to specify locations and plan movements. But you also need sensed information, represented by place cells, to associate sensory input with locations in the reference frame.

The mapping mechanisms in the neocortex are not an exact copy of ones in the old brain. Evidence suggests that the neocortex uses the same basic neural mechanisms, but it is different in several ways. It is as if nature stripped down the hippocampus and entorhinal cortex to a minimal form, made tens of thousands of copies, and arranged them side by side in cortical columns. That became the neocortex.

Grid cells and place cells in the old brain mostly track the location of one thing: the body. They know where the body is in its current environment. The neocortex, on the other hand, has about 150,000 copies of this circuit, one per cortical column. Therefore, the neocortex tracks thousands of locations simultaneously. For example, each small patch of your skin and each small patch of your retina has its own reference frame in the neocortex. Your five fingertips touching a cup are like five rats exploring a box.

Huge Maps in Tiny Spaces

So, what does a model in the brain look like? How does the neocortex stuff hundreds of models into each square millimeter? To understand how this works, let's go back to our paper-map analogy. Say I have a map of a town. I spread it out on a table and see that it is marked with rows and columns dividing it into one hundred squares. A1 is the top left and J10 is the bottom right. Printed in each square are things I might see in that part of town.

I take a pair of scissors and cut out each square, marking it with its grid coordinates: B6, G1, etc. I also mark each square with Town 1. I then do the same for nine more maps, each map representing a different town. I now have one thousand squares: one hundred map squares for each of ten towns. I shuffle the squares and put them in a stack. Although my stack contains ten complete maps, only one location can be seen at a time. Now someone blindfolds me and drops me off at a random location in one of the ten towns. Removing my blindfold, I look around. At first, I don't know where I am. Then I see I am standing in front of a fountain with a sculpture of a woman reading a book. I flip through my map squares, one at a time, until I see one showing this fountain. The map square is labeled Town 3, location D2. Now I know what town I am in and I know where I am in that town.

There are several things I can do next. For example, I can predict what I will see if I start walking. My current location is D2. If I walk east, I will be in D3. I search my stack of squares to find the square labeled Town 3, D3. It shows a playground. In this way I can predict what I will encounter if I move in a certain direction.

Perhaps I want to go to the town library. I can search my stack of squares until I see one showing a library in Town 3. That square is labeled G7. Given that I am at D2, I can calculate that I have to travel three squares east and five squares south to get to the library. I can take several different routes to get there. Using my map squares, one at a time, I can visualize what I will encounter along any particular route. I choose one that takes me past an ice cream shop.

Now consider a different scenario. After being dropped off at an unknown location and removing my blindfold, I see a coffee shop. But when I look through my stack of squares, I find five showing a similar-looking coffee shop. Two coffee shops are in one town, and the other three are in different towns. I could be in any of these five locations. What should I do? I can eliminate the ambiguity by moving. I look at the five squares where I might be, and then look up what I will see if I walk south from each of them. The answer is different for each of the five squares. To figure out where I am, I then physically walk south. What I find there eliminates my uncertainty. I now know where I am.

This way of using maps is different than how we typically use them. First, our stack of map squares contains all our maps. In this way, we use the stack to figure out both what town we are in and where we are in that town.

Second, if we are uncertain where we are, then we can determine our town and location by moving. This is what happens when you reach into a black box and touch an unknown object with one finger. With a single touch you probably can't determine what object you are feeling. You might have to move your finger one or more times to make that determination. By moving, you discover two things at the same time: the moment you recognize what object you are touching, you also know where your finger is on the object.

Finally, this system can scale to handle a large number of maps and do so quickly. In the paper-map analogy, I described looking at the map squares one at a time. This could take a lot of time if you had many maps. Neurons, however, use what is called associative memory. The details are not important here, but it allows neurons to search though all the map squares at once. Neurons take the same amount of time to search through a thousand maps as to search through one.

Maps in a Cortical Column

Now let's consider how maplike models are implemented by neurons in the neocortex. Our theory says that every cortical column can learn models of complete objects. Therefore, every column—every square millimeter of the neocortex—has its own set of map squares. How a cortical column does this is complicated, and we don't yet understand it completely, but we understand the basics.

Recall that a cortical column has multiple layers of neurons. Several of these layers are needed to create the map squares. Here is a simplified diagram to give you a flavor of what we think is happening in a cortical column.

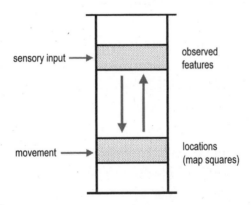

A model of a cortical column

This figure represents two layers of neurons (the shaded boxes) in one cortical column. Although a column is tiny, about one millimeter wide, each of these layers might have ten thousand neurons.

The upper layer receives the sensory input to the column. When an input arrives, it causes several hundred neurons to become active. In the paper-map analogy, the upper layer represents what you observe at some location, such as the fountain.

The bottom layer represents the current location in a reference frame. In the analogy, the lower layer represents a location—such as Town 3, D2—but doesn't represent what is observed there. It is like a blank square, labeled only with Town 3, location D2.

The two vertical arrows represent connections between the blank map squares (the lower layer) and what is seen at that location (the upper layer). The downward arrow is how an observed feature, such as the fountain, is associated with a particular location in a particular town. The upward arrow associates a particular location—Town 3, D2—with an observed feature. The upper layer is roughly equivalent to place cells and the lower layer is roughly equivalent to grid cells.

Learning a new object, such as a coffee cup, is mostly accomplished by learning the connections between the two layers, the vertical arrows. Put another way, an object such as a coffee cup is defined by a set of observed features (upper layer) associated with a set of locations on the cup (lower layer). If you know the feature, then you can determine the location. If you know the location, you can predict the feature.

The basic flow of information goes as follows: A sensory input arrives and is represented by the neurons in the upper layer. This invokes the location in the lower layer that is associated with the input. When movement occurs, such as moving a finger, then the lower layer changes to the expected new location, which causes a prediction of the next input in the upper layer.

If the original input is ambiguous, such as the coffee shop, then the network activates multiple locations in the lower layer—for example, all the locations where a coffee shop exists. This is what happens if you touch the rim of a coffee cup with one finger. Many objects have a rim, so you can't at first be certain what object you are touching. When you move, the lower layer changes all the possible locations, which then make multiple predictions in the upper layer. The next input will eliminate any locations that don't match.

We simulated this two-layer circuit in software using realistic assumptions for the number of neurons in each layer. Our simulations showed that not only can individual cortical columns learn models of objects, but each column can learn hundreds of them. The neural mechanism and simulations are described in our 2019 paper "Locations in the Neocortex: A Theory of Sensorimotor Object Recognition Using Cortical Grid Cells."

Orientation

There are other things a cortical column must do to learn models of objects. For example, there needs to be a representation of orientation. Say you know what town you are in and you know your location in that town. Now I ask you, "What will you see if you walk forward one block?" You would reply, "Which direction am I walking?" Knowing your location is not sufficient to predict what you will see when you walk; you also need to know which way you are facing, your orientation. Orientation is also required to predict what you will see from a particular location. For example, standing on a street corner, you might see a library when you face north and a playground when you face south.

There are neurons in the old brain called head direction cells. As their name suggests, these cells represent the direction an animal's head is facing. Head direction cells act like a compass, but they aren't tied to magnetic north. They are aligned to a room or environment. If you stand in a familiar room and then close your eyes, you retain a sense of which way you are facing. If you turn your body, while keeping your eyes closed, your sense of direction changes. This sense is created by your head direction cells. When you rotate your body, your head direction cells change to reflect your new orientation in the room.

Cortical columns must have cells that perform an equivalent function to head direction cells. We refer to them by the more generic term orientation cells. Imagine you are touching the lip of the coffee cup with your index finger. The actual impression on the finger depends on the orientation of the finger. You can, for example, keep your finger in the same location but rotate it around the point of contact. As you do, the sensation of the finger changes. Therefore, in order to predict its input, a cortical column must have a representation of orientation. For simplicity, I didn't show orientation cells and other details in the above diagram of a cortical column.

To summarize, we proposed that every cortical column learns models of objects. The columns do this using the same basic method that the old brain uses to learn models of environments. Therefore, we proposed that each cortical column has a set of cells equivalent to grid cells, another set equivalent to place cells, and another set equivalent to head direction cells, all of which were first discovered in parts of the old brain. We came to our hypothesis by logical deduction. In Chapter 7, I will list the growing experimental evidence that supports our proposal.

But first, we are going to turn our attention to the neocortex as a whole. Recall that each cortical column is small, about the width of a piece of thin spaghetti, and the neocortex is large, about the size of a dinner napkin. Therefore, there are about 150,000 columns in a human neocortex. Not all of the cortical columns are modeling objects. What the rest of the columns are doing is the topic of the next chapter.

CHAPTER 6

Concepts, Language, and High-Level Thinking

Our superior cognitive functions are what most distinguish us from our primate cousins. Our ability to see and hear is similar to a monkey's, but only humans use complex language, make complex tools such as computers, and are able to reason about concepts such as evolution, genetics, and democracy.

Vernon Mountcastle proposed that every column in the neocortex performs the same basic function. For this to be true, then, language and other high-level cognitive abilities are, at some fundamental level, the same as seeing, touching, and hearing. This is not obvious. Reading Shakespeare does not seem similar to picking up a coffee cup, but that is the implication of Mountcastle's proposal.

Mountcastle knew that cortical columns aren't completely identical. There are physical differences, for example, between columns that get input from your fingers and columns that understand language, but there are more similarities than differences. Therefore, Mountcastle deduced that there must be some basic function that underlies everything the neocortex does—not just perception, but all the things we think of as intelligence.

The idea that diverse abilities such as vision, touch, language, and philosophy are fundamentally the same is hard for many people to accept. Mountcastle didn't propose what the common function is, and it is hard to imagine what it could be, so it is easy to ignore his proposal or reject it outright. For example, linguists often describe language as different from all other cognitive abilities. If they embraced Mountcastle's proposal, they might look for the commonality between language and vision to better understand language. For me, this idea is too exciting to ignore, and I find that the empirical evidence overwhelmingly supports Mountcastle's proposal. Therefore, we are left with a fascinating puzzle: What kind of function, or algorithm, can create all aspects of human intelligence?

So far, I have described a theory of how cortical columns learn models of physical objects such as coffee cups, chairs, and smartphones. The theory says that cortical columns create reference frames for each observed object. Recall that a reference frame is like an invisible, three-dimensional grid surrounding and attached to something. The reference frame allows a cortical column to learn the locations of features that define the shape of an object.

In more abstract terms, we can think of reference frames as a way to organize any kind of knowledge. A reference frame for a coffee cup corresponds to a physical object that we can touch and see. However, reference frames can also be used to organize knowledge of things we can't directly sense.

Think of all the things you know that you haven't directly experienced. For example, if you have studied genetics, then you know about DNA molecules. You can visualize their double-helix shape, you know how they encode sequences of amino acids using the ATCG code of nucleotides, and you know how DNA molecules replicate by unzipping. Of course, nobody has ever directly seen or touched a DNA molecule. We can't because they are too small. To organize our knowledge of DNA molecules, we make pictures as if we could see them and models as if we could touch them. This allows us to store our knowledge of DNA molecules in reference frames—just like our knowledge of coffee cups.

We use this trick for much of what we know. For example, we know a lot about photons, and we know a lot about our galaxy, the Milky Way. Once again, we imagine these as if we could see and touch them, and therefore we can organize the facts we know about them using the same reference-frame mechanism that we use for everyday physical objects. But human knowledge extends to things that cannot be visualized. For example, we have knowledge about concepts such as democracy, human rights, and mathematics. We know many facts about these concepts, but we are unable to organize these facts in a way that resembles a three-dimensional object. You can't easily make an image of democracy.

But there must be some form of organization to conceptual knowledge. Concepts such as democracy and mathematics are not just a pile of facts. We are able to reason about them and make predictions about what will happen if we act one way or another. Our ability to do this tells us that knowledge of concepts must also be stored in reference frames. But these reference frames may not be easily equated to the reference frames we use for coffee cups and other physical objects. For example, it is possible that the reference frames that are most useful for certain concepts have more than three dimensions. We are not able to visualize spaces with more than three dimensions, but from a mathematical point of view they work the same way as spaces with three or fewer dimensions.

All Knowledge Is Stored in Reference Frames

The hypothesis I explore in this chapter is that the brain arranges all knowledge using reference frames, and that thinking is a form of moving. Thinking occurs when we activate successive locations in reference frames.

This hypothesis can be broken down into the following components.

1. Reference Frames Are Present Everywhere in the Neocortex

This premise states that every column in the neocortex has cells that create reference frames. I have proposed that the cells that do this are similar, but not identical, to the grid cells and place cells found in older parts of the brain.

2. Reference Frames Are Used to Model Everything We Know, Not Just Physical Objects

A column in the neocortex is just a bunch of neurons. A column doesn't "know" what its inputs represent, and it doesn't have any prior knowledge of what it is supposed to learn. A column is just a mechanism built of neurons that blindly tries to discover and model the structure of whatever is causing its inputs to change.

Earlier, I posited that brains first evolved reference frames to learn the structure of environments so that we could move about the world. Then our brains evolved to use the same mechanism to learn the structure of physical objects so that we could recognize and manipulate them. I am now proposing that our brains once again evolved to use that same mechanism to learn and represent the structure underlying conceptual objects, such as mathematics and democracy.

3. All Knowledge Is Stored at Locations Relative to Reference Frames

Reference frames are not an optional component of intelligence; they are the structure in which all information is stored in the brain. Every fact you know is paired with a location in a reference frame. To become an expert in a field such as history requires assigning historical facts to locations in an appropriate reference frame.

Organizing knowledge this way makes the facts actionable. Recall the analogy of a map. By placing facts about a town onto a grid-like reference frame, we can determine what actions are needed to achieve a goal, such as how to get to a particular restaurant. The uniform grid of the map makes the facts about the town actionable. This principle applies to all knowledge.

4. Thinking Is a Form of Movement

If everything we know is stored in reference frames, then to recall stored knowledge we have to activate the appropriate locations in the appropriate reference frames. Thinking occurs when the neurons invoke location after location in a reference frame, bringing to mind what was stored in each location. The succession of thoughts that we experience when thinking is analogous to the succession of sensations we experience when touching an object with a finger, or the succession of things we see when we walk about a town.

Reference frames are also the means for achieving goals. Just as a paper map allows you to figure out how to get from where you are to a desired new location, reference frames in the neocortex allow you to figure out the steps you should take to achieve more conceptual goals, such as solving an engineering problem or getting a promotion at work.

Although we mentioned these ideas about conceptual knowledge in some of our published research, they were not the focus, nor have we published papers directly on this topic. So, you could consider this chapter to be more speculative than earlier parts of the book, but I don't feel that way. Although there are many details we don't yet understand, I am confident that the overall framework—that concepts and thinking are based on reference frames—will withstand the test of time.

In the rest of this chapter, I will first describe a well-studied feature of the neocortex, its division into "what" regions and "where" regions. I use this discussion to show how cortical columns can perform markedly different functions by a simple change to their reference frames. Then I move on to more abstract and conceptual forms of intelligence. I present experimental evidence that supports the premises above and give examples of how the theory might relate to three topics: mathematics, politics, and language.

What and Where Pathways

Your brain has two vision systems. If you follow the optic nerve as it travels from the eye to the neocortex, you will see that it leads to two parallel vision systems, called the what visual pathway and the where visual pathway. The what pathway is a set of cortical regions that starts at the very back of the brain and moves around to the sides. The where pathway is a set of regions that also starts at the back of the brain but moves up toward the top.

The what and where vision systems were discovered over fifty years ago. Years later, scientists realized that similar, parallel pathways also exist for other senses. There are what and where regions for seeing, touching, and hearing.

What and where pathways have complementary roles. For example, if we disable the where visual pathway, then a person looking at an object can tell you *what* the object is, but they can't reach for the object. They know they are seeing a cup, for example, but, oddly, they can't say *where* the cup is. If we then switch it around and disable the what visual pathway, then the person can reach out and grab the object. They know *where* it is, but they cannot identify *what* it is. (At least not visually. When their hand touches the object, they can identify the object via touch.)

Columns in the what and where regions look similar. They have similar cell types, layers, and circuits. So why would they operate differently? What is the difference between a column in a what region and a column in a where region that leads to their different roles? You might be tempted to assume that there is some difference in how the two types of columns function. Perhaps where columns have a few additional types of neurons or different connections between layers. You might admit that what and where columns look similar but argue that there is probably some physical difference that we just haven't found yet. If you took this position, then you would be rejecting Mountcastle's proposal.

But it is not necessary to abandon Mountcastle's premise. We have proposed a simple explanation for why some columns are what columns and some are where columns. Cortical grid cells in what columns attach reference frames to objects. Cortical grid cells in where columns attach reference frames to your body.

If a where visual column could talk, it might say, "I have created a reference frame that is attached to the body. Using this reference frame, I look at a hand and know its location relative to the body. I then look at an object and know its location relative to the body. With these two locations, both in the body's reference frame, I can calculate how to move the hand to the object. I know where the object is and how to reach for it, but I can't identify it. I don't know what that object is."

If a what visual column could talk, it might say, "I have created a reference frame that is attached to an object. Using this reference frame, I can identify the object as a coffee cup. I know what the object is, but I don't know where it is." Working together, what and where columns allow us to identify objects, reach for them, and manipulate them.

Why would one column (column A) attach reference frames to an external object and another (column B) attach them to the body? It could be as simple as where the inputs to the column come from. If column A gets sensory input from an object, such as sensations from a finger touching a cup, it will automatically create a reference frame anchored to the object. If column B gets input from the body, such as neurons that detect the joint angles of the limbs, it will automatically create a reference frame anchored to the body.

In some ways, your body is just another object in the world. The neocortex uses the same basic method to model your body as it does to model objects such as coffee cups. However, unlike external objects, your body is always present. A significant portion of the neocortex—the where regions—is dedicated to modeling your body and the space around your body.

The idea that the brain contains maps of the body is not new. Neither is the idea that movement of the limbs requires body-centric reference frames. The point I want to make is that cortical columns, which look and operate similarly, can appear to perform different functions depending on what their reference frames are anchored to. Given this notion, it is not a big leap to see how reference frames can be applied to concepts.

Reference Frames for Concepts

Up to now in the book, I have described how the brain learns models of things that have a physical shape. Staplers, cell phones, DNA molecules, buildings, and your body all have physical presence. These are all things that we can directly sense or—as in the case of the DNA molecule—can imagine sensing.

However, much of what we know about the world can't be sensed directly and may not have any physical equivalent. For example, we can't reach out and touch concepts such as democracy or prime numbers, yet we know a lot about these things. How can cortical columns create models of things that we can't sense?

The trick is that reference frames don't have to be anchored to something physical. A reference frame for a concept such as democracy needs to be self-consistent, but it can exist relatively independent of everyday physical things. It is similar to how we can create maps for fictional lands. A map of a fictional land needs to be self-consistent, but it doesn't need to be located anywhere in particular relative to Earth.

The second trick is that reference frames for concepts do not have to have the same number or type of dimensions as reference frames for physical objects such as coffee cups. The locations of buildings in a town are best described in two dimensions. The shape of a coffee cup is best described in three dimensions. But all the abilities we get from a reference frame—such as determining the distance between two locations and calculating how to move from one location to another—are also present in reference frames with four or more dimensions.

If you have trouble understanding how something can have more than three dimensions, consider this analogy. Say I want to create a reference frame in which I can organize knowledge about all the people I know. One dimension I might use is age. I can arrange my acquaintances along this dimension by how old they are. Another metric might be where they live relative to me. This would require two more dimensions. Another dimension could be how often I see them, or how tall they are. I am up to five dimensions. This is just an analogy; these would not be the actual dimensions used by the neocortex. But I hope you can see how more than three dimensions could be useful.

It is likely that columns in the neocortex don't have a pre-conceived notion of what kind of reference frame they should use. When a column learns a model of something, part of the learning is discovering what is a good reference frame, including the number of dimensions.

Now, I will review the empirical evidence that supports the four premises I listed above. This is an area where there isn't a lot of experimental evidence, but there is some, and it is growing.

Method of Loci

A well-known trick for remembering a list of items, known as the method of loci or sometimes the memory palace, is to imagine placing the items you want to remember at different locations in your house. To recall the list of items, you imagine walking through your house, which brings back the memory of each item one at a time. The success of this memory trick tells us that recalling things is easier when they are assigned to locations in a familiar reference frame. In this case, the reference frame is the mental map of your house. Notice that the act of recalling is achieved by moving. You are not physically moving your body, but mentally moving through your house.

The method of loci supports two of the above premises: information is stored in reference frames and retrieval of information is a form of movement. The method is useful for quickly memorizing a list of items, such as a random set of nouns. It works because it assigns the items to a previously learned reference frame (your house) and uses previously learned movements (how you typically move through your house). However, most of the time when you learn, your brain creates new reference frames. We will see an example of that next.

Studies of Humans Using fMRI

fMRI is a technology for looking into a live brain and seeing which parts are most active. You have probably seen fMRI images: they show an outline of a brain with some parts colored yellow and red, indicating where the most energy was being consumed when the image was taken. fMRI is usually used with a human subject because the process requires lying perfectly still inside a narrow tube in a big noisy machine while doing a specific mental task. Often, the subject will be looking at a computer screen while following the verbal instructions of a researcher.

The invention of fMRI has been a boon for certain types of research, but for the kind of research we do it is generally not too useful. Our research on neocortical theory relies on knowing which individual neurons are active at any point in time, and the active neurons change several times a second. There are experimental techniques that provide this kind of data, but fMRI does not have the spatial and temporal precision we typically need. fMRI measures the average activity of many neurons and can't detect activity that lasts less than about a second.

Therefore, we were surprised and elated to learn of a clever fMRI experiment performed by Christian Doeller, Caswell Barry, and Neil Burgess that showed that grid cells exist in the neocortex. The details are complicated, but the researchers realized that grid cells might exhibit a signature that could be detected using fMRI. They first had to verify that their technique worked, so they looked at the entorhinal cortex, where grid cells are known to exist. They had human subjects perform a navigation task of moving around a virtual world on a computer screen and, using fMRI, they were able to detect the presence of grid-cell activity while the subjects were performing the task. Then they turned their focus to the neocortex. They used their fMRI technique to look at frontal areas of the neocortex while the subject performed the same navigation task. They found the same signature, strongly suggesting that grid cells also exist in at least some parts of the neocortex.

Another team of scientists, Alexandra Constantinescu, Jill O'Reilly, and Timothy Behrens, used the new fMRI technique for a different task. The subjects were shown images of birds. The birds differed by the length of their necks and the length of their legs. The subjects were asked to perform various mental imagery tasks related to the birds, such as to imagine a new bird that combined features of two previously seen birds. Not only did the experiments show that grid cells are present in frontal areas of the neocortex, but the researchers found evidence that the neocortex stored the bird images in a maplike reference frame—one dimension represented neck length and another represented leg length. The research team further showed that when the subjects thought about birds, they were mentally "moving" through the map of birds in the same way you can mentally move through the map of your house. Again, the details of this experiment are complex, but the fMRI data suggest that this part of the neocortex used grid-cell-like neurons to learn about birds. The subjects who participated in this experiment had no notion that this was happening, but the imaging data were clear.

The method of loci uses a previously learned map, the map of your house, to store items for later recall. In the bird example, the neocortex created a new map, a map that was suited for the task of remembering birds with different necks and legs. In both examples, the process of storing items in a reference frame and recalling them via "movement" is the same.

If all knowledge is stored this way, then what we commonly call thinking is actually moving through a space, through a reference frame. Your current thought, the thing that is in your head at any moment, is determined by the current location in the reference frame. As the location changes, the items stored at each location are recalled one at a time. Our thoughts are continually changing, but they are not random. What we think next depends on which direction we mentally move through a reference frame, in the same way that what we see next in a town depends on which direction we move from our current location.

The reference frame needed to learn a coffee cup is perhaps obvious: it is the three-dimensional space around the cup. The reference frame learned in the fMRI experiment about birds is perhaps a bit less obvious. But the bird reference frame is still related to the physical attributes of birds, such as legs and necks. But what kind of reference frame should the brain use for concepts like economics or ecology? There may be multiple reference frames that work, although some may be better than others.

This is one reason that learning conceptual knowledge can be difficult. If I give you ten historical events related to democracy, how should you arrange them? One teacher might show the events arranged on a timeline. A timeline is a one-dimensional reference frame. It is useful for assessing the temporal order of events and which events might be causally related by temporal proximity. Another teacher might arrange the same historical events geographically on a map of the world. A map reference frame suggests different ways of thinking about the same events, such as which events might be causally related by spatial proximity to each other, or by proximity to oceans, deserts, or mountains. Timelines and geography are both valid ways of organizing historical events, yet they lead to different ways of thinking about history. They might lead to different conclusions and different predictions. The best structure for learning about democracy might require an entirely new map, a map with multiple abstract dimensions that correspond to fairness or rights. I am not suggesting that "fairness" or "rights" are actual dimensions used by the brain. My point is that becoming an expert in a field of study requires discovering a good framework to represent the associated data and facts. There may not be a *correct* reference frame, and two individuals might arrange the facts differently. Discovering a useful reference frame is the most difficult part of learning, even though most of the time we are not consciously aware of it. I will illustrate this idea with the three examples I mentioned earlier: mathematics, politics, and language.

Mathematics

Say you are a mathematician and you want to prove the OMG conjecture (OMG is not a real conjecture). A conjecture is a mathematical statement that is believed to be true but that has not been proven. To prove a conjecture, you start with something that is known to be true. Then you apply a series of mathematical operations. If, through this process, you arrive at a statement that is the conjecture, then you have succeeded in proving it. Typically, there will be a series of intermediate results. For example, starting from A, prove B. From B, prove C. And finally, from C, prove OMG. Let's say, A, B, C, and the final OMG are equations. To get from equation to equation, you have to perform one or more mathematical operations.

Now let's suppose that the various equations are represented in your neocortex in a reference frame. Mathematical operations, such as multiply or divide, are movements that take you to different locations in this reference frame. Performing a series of operations moves you to a new location, a new equation. If you can determine a set of operations—movements through equation space—that gets you from A to OMG, then you have succeeded in proving OMG.

Solving complex problems, like a mathematical conjecture, requires a lot of training. When learning a new field, your brain is not just storing facts. For mathematics, the brain must discover useful reference frames in which to store equations and numbers, and it must learn how mathematical behaviors, such as operations and transforms, move to new locations within the reference frames.

To a mathematician, equations are familiar objects, similar to how you and I see a smartphone or a bicycle. When mathematicians see a new equation, they recognize it as similar to previous equations they have worked with, and this immediately suggests how they can manipulate the new equation to achieve certain results. It is the same process we go through if we see a new smartphone. We recognize the phone is similar to other phones we have used and that suggests how we could manipulate the new phone to achieve a desired outcome.

However, if you are not trained in mathematics, then equations and other mathematical notations will appear as meaningless scribbles. You may even recognize an equation as one you have seen before, but without a reference frame, you will have no idea how to manipulate it to solve a problem. You can be lost in math space, in the same way you can be lost in the woods without a map.

Mathematicians manipulating equations, explorers traveling through a forest, and fingers touching coffee cups all need maplike reference frames to know where they are and what movements they need to perform to get where they want to be. The same basic algorithm underlies these and countless other activities we perform.

Politics

The above mathematical example is completely abstract, but the process is the same for any problem that isn't obviously physical.

For example, say a politician wants to get a new law enacted. They have a first draft of the law written, but there are multiple steps required to get to the end goal of enactment. There are political obstacles along the way, so the politician thinks about all the different actions they might take. An expert politician knows what will likely happen if they hold a press conference or force a referendum or write a policy paper or offer to trade support for another bill. A skilled politician has learned a reference frame for politics. Part of the reference frame is how political actions change locations in the reference frame, and the politician imagines what will happen if they do these things. Their goal is to find a series of actions that will lead them to the desired result: enacting the new law.

A politician and a mathematician are not aware they are using reference frames to organize their knowledge, just as you and I are not aware that we use reference frames to understand smartphones and staplers. We don't go around asking, "Can someone suggest a reference frame for organizing these facts?" What we do say is, "I need help. I don't understand how to solve this problem." Or "I'm confused. Can you show me how to use this thing?" Or "I'm lost. Can you show me how to get to the cafeteria?" These are the questions we ask when we are not able to assign a reference frame to the facts in front of us.

Language

Language is arguably the most important cognitive ability that distinguishes humans from all other animals. Without the ability to share knowledge and experiences via language, most of modern society would not be possible.

Although many volumes have been written about language, I am not aware of any attempts to explain how language is created by the neural circuits observed in the brain. Linguists don't usually venture into neuroscience, and although some neuroscientists study brain regions related to language, they have been unable to propose detailed theories of how the brain creates and understands language.

There is an ongoing debate as to whether language is fundamentally different from other cognitive abilities. Linguists tend to think so. They describe language as a unique capability, unlike anything else we do. If this were true, then the parts of the brain that create and understand language should look different. Here, the neuroscience is equivocal.

There are two modest-size regions of the neocortex that are said to be responsible for language. Wernicke's area is thought to be responsible for language comprehension, and Broca's area is thought to be responsible for language production. This is a bit of a simplification. First, there is disagreement over the exact location and extent of these regions. Second, the functions of Wernicke's and Broca's areas are not neatly differentiated into comprehension and production; they overlap a bit. Finally, and this should be obvious, language can't be isolated to two small regions of the neocortex. We use spoken language, written language, and sign language. Wernicke's and Broca's areas don't get input directly from the sensors, so the comprehension of language must rely on auditory and visual regions, and the production of language must rely on different motor abilities. Large areas of the neocortex are needed to create and understand language. Wernicke's and Broca's areas play a key role, but it is wrong to think of them as creating language in isolation.

A surprising thing about language, and one that suggests that language may be different than other cognitive functions, is that Broca's and Wernicke's areas are only on the left side of the brain. The equivalent areas on the right side are only marginally implicated in language. Almost everything else the neocortex does occurs on both sides of the brain. The unique asymmetry of language suggests there is something different about Broca's and Wernicke's areas.

Why language only occurs on the left side of the brain might have a simple explanation. One proposal is that language requires fast processing, and neurons in most of the neocortex are too slow to process language. The neurons in Wernicke's and Broca's areas are known to have extra insulation (called myelin) that allows them to run faster and keep up with the demands of language. There are other noticeable differences with the rest of the neocortex. For example, it has been reported that the number and density of synapses is greater in the language regions compared to their equivalents on the right side of the brain. But having more synapses doesn't mean that the language areas perform a different function; it could just mean that these areas have learned more things.

Although there are some differences, the anatomy of Wernicke's and Broca's areas is, once again, similar to other areas of the neocortex. The facts that we have today suggest that, while these language areas are somewhat different, perhaps in subtle ways, the overall structure of layers, connectivity, and cell types is similar to the rest of the neocortex. Therefore, the majority of the mechanisms underlying language are likely shared with other parts of cognition and perception. This should be our working assumption until proven otherwise. Thus we can ask, How could the modeling capabilities of a cortical column, including reference frames, provide a substrate for language?

According to linguists, one of the defining attributes of language is its nested structure. For example, sentences are composed of phrases, phrases are composed of words, and words are composed of letters. Recursion, the ability to repeatedly apply a rule, is another defining attribute. Recursion allows sentences to be constructed with almost unlimited complexity. For example, the simple sentence "Tom asked for more tea" can be extended to "Tom, who works at the auto shop, asked for more tea," which can be extended to "Tom, who works at the auto shop, the one by the thrift store, asked for more tea." The exact definition of recursion as it relates to language is debated, but the general idea is not hard to understand. Sentences can be composed of phrases, which can be composed of other phrases, and so on. It has long been argued that nested structure and recursion are key attributes of language.

However, nested and recursive structure is not unique to language. In fact, everything in the world is composed this way. Take my coffee cup with the Numenta logo printed on the side. The cup has a nested structure: it consists of a cylinder, a handle, and a logo. The logo consists of a graphic and a word. The graphic is made up of circles and lines, while the word "Numenta" is composed of syllables, and the syllables are themselves made of letters. Objects can also have recursive structure. For example, imagine that the Numenta logo included a picture of a coffee cup, on which was printed a picture of the Numenta logo, which had a picture of a coffee cup, etc.

Early on in our research, we realized that each cortical column had to be able to learn nested and recursive structure. This was a constraint necessary to learn the structure of physical things like coffee cups and to learn the structure of conceptual things like mathematics and language. Any theories we came up with had to explain how columns do this.

Imagine that sometime in the past you learned what a coffee cup looks like, and sometime in the past you learned what the Numenta logo looks like. But you had never seen the logo on a coffee cup. Now I show you a new coffee cup with the logo on the side. You can learn the new combined object quickly, usually with just one or two glances. Notice that you don't need to relearn the logo or the cup. Everything we know about cups and the logo is immediately included as part of the new object.

How does this happen? Within a cortical column, the previously learned coffee cup is defined by a reference frame. The previously learned logo is also defined by a reference frame. To learn the coffee cup with the logo, the column creates a new reference frame, in which it stores two things: a link to the reference frame of the previously learned cup and a link to the reference frame of the previously learned logo. The brain can do this rapidly, with just a few additional synapses. This is a bit like using hyperlinks in a text document. Imagine I wrote a short essay about Abraham Lincoln and I mention that he gave a famous speech called the Gettysburg Address. By turning the words "Gettysburg Address" into a link to the full speech, I can include all the details of the speech as part of my essay without having to retype it.

Earlier I said that cortical columns store features at locations in reference frames. The word "feature" is a bit vague. I will now be more precise. Cortical columns create reference frames for every object they know. Reference frames are then populated with links to other reference frames. The brain models the world using reference frames that are populated with reference frames; it is reference frames all the way down. In our 2019 "Frameworks" paper, we proposed how neurons might do this.

We have a long way to go to fully understand everything the neocortex does. But the idea that every column models objects using reference frames is, as far as we know, consistent with the needs of language. Perhaps further down the road we will find a need for some special language circuits. But for now, that is not the case.

Expertise

By now I have introduced four uses for reference frames, one in the old brain and three in the neocortex. Reference frames in the old brain learn maps of environments. Reference frames in the what columns of the neocortex learn maps of physical objects. Reference frames in the where columns of the neocortex learn maps of the space around our body. And, finally, reference frames in the non-sensory columns of the neocortex learn maps of concepts.

To be an expert in any domain requires having a good reference frame, a good map. Two people observing the same physical object will likely end up with similar maps. For example, it is hard to imagine how the brains of two people observing the same chair would arrange its features differently. But when thinking about concepts, two people starting with the same facts might end up with different reference frames. Recall the example of a list of historical facts. One person might arrange the facts on a timeline, and another might arrange them on a map. The same facts can lead to different models and different worldviews.

Being an expert is mostly about finding a good reference frame to arrange facts and observations. Albert Einstein started with the same facts as his contemporaries. However, he found a better way to arrange them, a better reference frame, that permitted him to see analogies and make predictions that were surprising. What is most fascinating about Einstein's discoveries related to special relativity is that the reference frames he used to make them were everyday objects. He thought about trains, people, and flashlights. He started with the empirical observations of scientists, such as the absolute speed of light, and used everyday reference frames to deduce the equations of special relativity. Because of this, almost anyone can follow his logic and understand how he made his discoveries. In contrast, Einstein's general theory of relativity required reference frames based on mathematical concepts called field equations, which are not easily related to everyday objects. Einstein found this much harder to understand, as does pretty much everyone else.

In 1978, when Vernon Mountcastle proposed that there was a common algorithm underlying all perception and cognition, it was hard to imagine what algorithm could be powerful enough and general enough to fit the requirement. It was hard to imagine a single process that could explain everything we think of as intelligence, from basic sensory perception to the highest and most admired forms of intellectual ability. It is now clear to me that the common cortical algorithm is based on reference frames. Reference frames provide the substrate for learning the structure of the world, where things are, and how they move and change. Reference frames can do this not just for the physical objects that we can directly sense, but also for objects we cannot see or feel and even for concepts that have no physical form.

Your brain has 150,000 cortical columns. Each column is a learning machine. Each column learns a predictive model of its inputs by observing how they change over time. Columns don't know what they are learning; they don't know what their models represent. The entire enterprise and the resultant models are built on reference frames. The correct reference frame to understand how the brain works is reference frames.

CHAPTER 7

The Thousand Brains
Theory of Intelligence

From its inception, Numenta's goal was to develop a broad theory of how the neocortex works. Neuroscientists were publishing thousands of papers a year covering every detail of the brain, but there was a lack of systemic theories that tied the details together. We decided to first focus on understanding a single cortical column. We knew cortical columns were physically complex and therefore must do something complex. It didn't make sense to ask why columns are connected to each other in the messy, somewhat hierarchical way I showed in Chapter 2 if we didn't know what a single column did. That would be like asking how societies work before knowing anything about people.

Now we know a lot about what cortical columns do. We know that each column is a sensory-motor system. We know that each column can learn models of hundreds of objects, and that the models are based on reference frames. Once we understood that columns did these things, it became clear that the neocortex, as a whole, worked differently than was previously thought. We call this new perspective the Thousand Brains Theory of Intelligence.

Before I explain what the Thousand Brains Theory is, it will be helpful to know what it is replacing.

The Existing View of the Neocortex

Today, the most common way of thinking about the neocortex is like a flowchart. Information from the senses is processed step-by-step as it passes from one region of the neocortex to the next. Scientists refer to this as a hierarchy of feature detectors. It is most often described in terms of vision, and goes like this: Each cell in the retina detects the presence of light in a small part of an image. The cells in the retina then project to the neocortex. The first region in the neocortex that receives this input is called region V1. Each neuron in region V1 gets input from only a small part of the retina. It is as if they were looking at the world through a straw.

These facts suggest that columns in region V1 cannot recognize complete objects. Therefore, V1's role is limited to detecting small visual features such as lines or edges in a local part of an image. Then the V1 neurons pass these features on to other regions of the neocortex. The next visual region, called V2, combines the simple features from region V1 into more complex features, such as corners or arcs. This process is repeated a couple more times in a couple more regions until neurons respond to complete objects. It is presumed that a similar process—going from simple features to complex features to complete objects—is also occurring with touch and hearing. This view of the neocortex as a hierarchy of feature detectors has been the dominant theory for fifty years.

The biggest problem with this theory is that it treats vision as a static process, like taking a picture. But vision is not like that. About three times a second our eyes make quick saccadic movements. The inputs from the eyes to the brain completely change with every saccade. The visual inputs also change when we walk forward or turn our head left and right. The hierarchy of features theory ignores these changes. It treats vision as if the goal is to take one picture at a time and label it. But even casual observation will tell you that vision is an interactive process, dependent on movement. For example, to learn what a new object looks like, we hold it in our hand, rotating it this way and that, to see what it looks like from different angles. Only by moving can we learn a model of the object.

One reason many people have ignored the dynamic aspect of vision is that we can sometimes recognize an image without moving our eyes, such as a picture briefly flashed on a display—but that is an exception, not the rule. Normal vision is an active sensory-motor process, not a static process.

The essential role of movement is more obvious with touch and hearing. If someone places an object onto your open hand, you cannot identify it unless you move your fingers. Similarly, hearing is always dynamic. Not only are auditory objects, such as spoken words, defined by sounds changing over time, but as we listen we move our head to actively modify what we hear. It is not clear how the hierarchy of features theory even applies to touch or hearing. With vision, you can at least imagine that the brain is processing a picture-like image, but with touch and hearing there is nothing equivalent.

There are numerous additional observations that suggest the hierarchy of features theory needs modification. Here are several, all related to vision:

- The first and second visual regions, V1 and V2, are some of the largest in the human neocortex. They are substantially larger in area than other visual regions, where complete objects are supposedly recognized. Why would detecting small features, which are limited in number, require a larger fraction of the brain than recognizing complete objects, of which there are many? In some mammals, such as the mouse, this imbalance is worse. Region V1 in the mouse occupies a large portion of the entire mouse neocortex. Other visual regions in the mouse are tiny in comparison. It is as if almost all of mouse vision occurs in region V1.

- The feature-detecting neurons in V1 were discovered when researchers projected images in front of the eyes of anesthetized animals while simultaneously recording the activity of neurons in V1. They found neurons that became active to simple features, such as an edge, in a small part of the image. Because the neurons only responded to simple features in a small area, they assumed that complete objects must be recognized elsewhere. This led to the hierarchical features model. But in these experiments, most of the neurons in V1 did not respond to anything obvious—they might emit a spike now and then, or they might spike continuously for a while then stop. The majority of neurons couldn't be explained with the hierarchy of features theory, so they were mostly ignored. However, all the unaccounted-for neurons in V1 must be doing something important that isn't feature detection.

- When the eyes saccade from one fixation point to another, some of the neurons in regions V1 and V2 do something remarkable. They seem to know what they will be seeing before the eyes have stopped moving. These neurons become active as if they can see the new input, but the input hasn't yet arrived. Scientists who discovered this were surprised. It implied that neurons in regions V1 and V2 had access to knowledge about the entire object being seen and not just a small part of it.

- There are more photoreceptors in the center of the retina than at the periphery. If we think of the eye as a camera, then it is one with a severe fish-eye lens. There are also parts of the retina that have no photoreceptors, for example, the blind spot where the optic nerve exits the eye and where blood vessels cross the retina. Consequentially, the input to the neocortex is not like a photograph. It is a highly distorted and incomplete quilt of image patches. Yet we are unaware of the distortions and missing pieces; our perception of the world is uniform and complete. The hierarchy of features theory can't explain how this happens. This problem is called the binding problem or the sensor-fusion problem. More generally, the binding problem asks how inputs from different senses, which are scattered all over the neocortex with all sorts of distortions, are combined into the singular non-distorted perception we all experience.

- As I pointed out in Chapter 1, although some of the connections between regions of the neocortex appear hierarchical, like a step-by-step flowchart, the majority do not. For example, there are connections between low-level visual regions and low-level touch regions. These connections do not make sense in the hierarchy of features theory.

- Although the hierarchy of features theory might explain how the neocortex recognizes an image, it provides no insight into how we learn the three-dimensional structure of objects, how objects are composed of other objects, and how objects change and behave over time. It doesn't explain how we can imagine what an object will look like if rotated or distorted.

With all these inconsistencies and shortcomings, you might be wondering why the hierarchy of features theory is still widely held. There are several reasons. First, it fits a lot of data, especially data collected a long time ago. Second, the problems with the theory accumulated slowly over time, making it easy to dismiss each new problem as small. Third, it is the best theory we have, and, without something to replace it, we stick with it. Finally, as I will argue shortly, it isn't completely wrong—it just needs a major upgrade.

The New View of the Neocortex

Our proposal of reference frames in cortical columns suggests a different way of thinking about how the neocortex works. It says that all cortical columns, even in low-level sensory regions, are capable of learning and recognizing complete objects. A column that senses only a small part of an object can learn a model of the entire object by integrating its inputs over time, in the same way that you and I learn a new town by visiting one location after another. Therefore, a hierarchy of cortical regions is not strictly needed to learn models of objects. Our theory explains how a mouse, with a mostly one-level visual system, can see and recognize objects in the world.

The neocortex has many models of any particular object. The models are in different columns. They are not identical, but complementary. For example, a column getting tactile input from a fingertip could learn a model of a cell phone that includes its shape, the textures of its surfaces, and how its buttons move when pressed. A column getting visual input from the retina could learn a model of the phone that also includes its shape, but, unlike the fingertip column, its model can include the color of different parts of the phone and how visual icons on the screen change as you use it. A visual column cannot learn the detent of the power switch and a tactile column cannot learn how icons change on the display.

Any individual cortical column cannot learn a model of every object in the world. That would be impossible. For one, there is a physical limit to how many objects an individual column can learn. We don't know yet what that capacity is, but our simulations suggest that a single column can learn hundreds of complex objects. This is much smaller than the number of things you know. Also, what a column learns is limited by its inputs. For example, a tactile column can't learn models of clouds and a visual column can't learn melodies.

Even within a single sensory modality, such as vision, columns get different types of input and will learn different types of models. For example, there are some vision columns that get color input and others that get black-and-white input. In another example, columns in regions V1 and V2 both get input from the retina. A column in region V1 gets input from a very small area of the retina, as if it is looking at the world through a narrow straw. A column in V2 gets input from a larger area of the retina, as if it is looking at the world through a wider straw, but the image is fuzzier. Now imagine you are looking at text in the smallest font that you can read. Our theory suggests that only columns in region V1 can recognize letters and words in the smallest font. The image seen by V2 is too fuzzy. As we increase the font size, then V2 and V1 can both recognize the text. If the font gets larger still, then it gets harder for V1 to recognize the text, but V2 is still able to do so. Therefore, columns in regions V1 and V2 might both learn models of objects, such as letters and words, but the models differ by scale.

Where Is Knowledge Stored in the Brain?

Knowledge in the brain is distributed. Nothing we know is stored in one place, such as one cell or one column. Nor is anything stored everywhere, like in a hologram. Knowledge of something is distributed in thousands of columns, but these are a small subset of all the columns.

Consider again our coffee cup. Where is knowledge about the coffee cup stored in the brain? There are many cortical columns in the visual regions that receive input from the retina. Each column that is seeing a part of the cup learns a model of the cup and tries to recognize it. Similarly, if you grasp the cup in your hands, then dozens to hundreds of models in the tactile regions of the neocortex become active. There isn't a single model of coffee cups. What you know about coffee cups exists in thousands of models, in thousands of columns—but, still, only in a fraction of all the columns in the neocortex. This is why we call it the Thousand Brains Theory: knowledge of any particular item is distributed among thousands of complementary models.

Here is an analogy. Say we have a city with one hundred thousand citizens. The city has a set of pipes, pumps, tanks, and filters to deliver clean water to every household. The water system needs maintenance to stay in good working order. Where does the knowledge reside for how to maintain the water system? It would be unwise to have only one person know this, and it would be impractical for every citizen to know it. The solution is to distribute the knowledge among many people, but not too many. In this case, let's say the water department has fifty employees. Continuing with this analogy, say there are one hundred parts of the water system—that is, one hundred pumps, valves, tanks, etc.—and each of the fifty workers in the water department knows how to maintain and repair a different, but overlapping, set of twenty parts.

OK then, where is knowledge of the water system stored? Each of the hundred parts is known by about ten different people. If half the workers called in sick one day, it is highly likely that there would still be five or so people available to repair any particular part. Each employee can maintain and fix 20 percent of the system on their own, with no supervision. Knowledge of how to maintain and repair the water system is distributed among a small number of the populace, and the knowledge is robust to a large loss of employees.

Notice that the water department might have some hierarchy of control, but it would be unwise to prevent any autonomy or to assign any piece of knowledge to just one or two people. Complex systems work best when knowledge and actions are distributed among many, but not too many, elements.

Everything in the brain works this way. For example, a neuron never depends on a single synapse. Instead, it might use thirty synapses to recognize a pattern. Even if ten of those synapses fail, the neuron will still recognize the pattern. A network of neurons is never dependent on a single cell. In the simulated networks we create, even the loss of 30 percent of the neurons usually has only a marginal effect on the performance of the network. Similarly, the neocortex is not dependent on a single cortical column. The brain continues to function even if a stroke or trauma wipes out thousands of columns.

Therefore, we should not be surprised that the brain does not rely on one model of anything. Our knowledge of something is distributed among thousands of cortical columns. The columns are not redundant, and they are not exact copies of each other. Most importantly, each column is a complete sensory-motor system, just as each water department worker is able to independently fix some portion of the water infrastructure.

The Solution to the Binding Problem

Why do we have a singular perception if we have thousands of models? When we hold and look at a coffee cup, why does the cup feel like one thing and not thousands of things? If we place the cup on a table and it makes a sound, how does the sound get united with the image and feel of the coffee cup? In other words, how do our sensory inputs get bound into a singular percept? Scientists have long assumed that the varied inputs to the neocortex must converge onto a single place in the brain where something like a coffee cup is perceived. This assumption is part of the hierarchy of features theory. However, the connections in the neocortex don't look like this. Instead of converging onto one location, the connections go in every direction. This is one of the reasons why the binding problem is considered a mystery, but we have proposed an answer: columns vote. Your perception is the consensus the columns reach by voting.

Let's go back to the paper map analogy. Recall that you have a set of maps for different towns. The maps are cut into little squares and mixed together. You are dropped off in an unknown location and see a coffee shop. If you find similar-looking coffee shops on multiple map squares, then you can't know where you are. If coffee shops exist in four different towns, then you know you must be in one of four towns, but you can't tell which one.

Now let's pretend there are four more people just like you. They also have maps of the towns and they are dropped off in the same town as you but in different, random locations. Like you, they don't know what town they are in or where they are. They take their blindfolds off and look around. One person sees a library and, after looking at his map squares, finds libraries in six different towns. Another person sees a rose garden and finds rose gardens in three different towns. The other two people do the same. Nobody knows what town they are in, but they all have a list of possible towns. Now everyone votes. All five of you have an app on your phone that lists the towns and locations you might be in. Everyone gets to see everyone else's list. Only Town 9 is on everyone's list; therefore, everyone now knows they are in Town 9. By comparing your lists of possible towns, and keeping only the towns on everybody's list, you all instantly know where you are. We call this process voting.

In this example, the five people are like five fingertips touching different locations on an object. Individually they can't determine what object they are touching, but together they can. If you touch something with only one finger, then you have to move it to recognize the object. But if you grasp the object with your entire hand, then you can usually recognize the object at once. In almost all cases, using five fingers will require less movement than using one. Similarly, if you look at an object through a straw, you have to move the straw to recognize the object. But if you view it with your entire eye, you can usually recognize it without moving.

Continuing with the analogy, imagine that, of the five people dropped off in the town, one person can only hear. That person's map squares are marked with the sounds they should hear at each location. When they hear a fountain, or birds in trees, or music from a cantina, they find the map squares where those sounds might be heard. Similarly, let's say two people can only touch things. Their maps are marked with the tactile sensations they expect to feel at different locations. Finally, two people can only see. Their map squares are marked with what they can expect to see at each location. We now have five people with three different types of sensors: vision, touch, and sound. All five people sense something, but they can't determine where they are, so they vote. The voting mechanism works identically as I described before. They only need to agree on the town—none of the other details matter. Voting works across sensory modalities.

Notice that you need to know little about the other people. You don't need to know what senses they have or how many maps they have. You don't need to know if their maps have more or fewer squares than your maps, or if the squares represent larger or smaller areas. You don't need to know how they move. Perhaps some people can hop over squares and others can only move diagonally. None of these details matter. The only requirement is that everyone can share their list of possible towns. Voting among cortical columns solves the binding problem. It allows the brain to unite numerous types of sensory input into a single representation of what is being sensed.

There is one more twist to voting. When you grasp an object in your hand, we believe the tactile columns representing your fingers share another piece of information—their relative position to each other, which makes it easier to figure out what they are touching. Imagine our five explorers are dropped into an unknown town. It is possible, indeed likely, that they see five things that exist in many towns, such as two coffee shops, a library, a park, and a fountain. Voting will eliminate any possible towns that don't have all these features, but the explorers still won't know for certain where they are, as several towns have all five features. However, if the five explorers know their relative position to each other, then they can eliminate any towns that don't have the five features in that particular arrangement. We suspect that information about relative position is also shared among some cortical columns.

How Is Voting Accomplished in the Brain?

Recall that most of the connections in a cortical column go up and down between the layers, largely staying within the bounds of the column. There are a few well-known exceptions to this rule.

Cells in some layers send axons long distances within the neocortex. They might send their axons from one side of the brain to the other, for example, between the areas representing the left and right hands. Or they might send their axons from V1, the primary visual region, to A1, the primary auditory region. We propose that these cells with long-distance connections are voting.

It only makes sense for certain cells to vote. Most of the cells in a column don't represent the kind of information that columns could vote on. For example, the sensory input to one column differs from the sensory input to other columns, and therefore cells that receive these inputs do not project to other columns. But cells that represent what object is being sensed can vote and will project broadly.

The basic idea of how columns can vote is not complicated. Using its long-range connections, a column broadcasts what it thinks it is observing. Often a column will be uncertain, in which case its neurons will send multiple possibilities at the same time. Simultaneously, the column receives projections from other columns representing their guesses. The most common guesses suppress the least common ones until the entire network settles on one answer. Surprisingly, a column doesn't need to send its vote to every other column. The voting mechanism works well even if the long-range axons connect to a small, randomly chosen subset of other columns. Voting also requires a learning phase. In our published papers, we described software simulations that show how learning occurs and how voting happens quickly and reliably.

Stability of Perception

Column voting solves another mystery of the brain: Why does our perception of the world seem stable when the inputs to the brain are changing? When our eyes saccade, the input to the neocortex changes with each eye movement, and therefore the active neurons must change too. Yet our visual perception is stable; the world does not appear to be jumping about as our eyes move. Most of the time, we are completely unaware that our eyes are moving at all. A similar stability of perception occurs with touch. Imagine a coffee cup is on your desk and you are grasping it with your hand. You perceive the cup. Now you mindlessly run your fingers over the cup. As you do this, the inputs to the neocortex change, but your perception is that the cup is stable. You do not think the cup is changing or moving.

So why is our perception stable, and why are we unaware of the changing inputs from our skin and eyes? Recognizing an object means the columns voted and now agree on what object they are sensing. The voting neurons in each column form a stable pattern that represents the object and where it is relative to you. The activity of the voting neurons does not change as you move your eyes and fingers, as long as they are sensing the same object. The other neurons in each column change with movement, but the voting neurons, the ones that represent the object, do not.

If you could look down on the neocortex, you would see a stable pattern of activity in one layer of cells. The stability would span large areas, covering thousands of columns. These are the voting neurons. The activity of the cells in other layers would be rapidly changing on a column-by-column basis. What we perceive is based on the stable voting neurons. The information from these neurons is spread broadly to other areas of the brain, where it can be turned into language or stored in short-term memory. We are not consciously aware of the changing activity within each column, as it stays within the column and is not accessible to other parts of the brain.

To stop seizures, doctors will sometimes cut the connections between the left and right sides of the neocortex. After surgery, these patients act as if they have two brains. Experiments clearly show that the two sides of the brain have different thoughts and reach different conclusions. Column voting can explain why. The connections between the left and right neocortex are used for voting. When they are cut, there is no longer a way for the two sides to vote, so they reach independent conclusions.

The number of voting neurons active at any time is small. If you were a scientist looking at the neurons responsible for voting, you might see 98 percent of the cells being silent and 2 percent continuously firing. The activity of the other cells in cortical columns would be changing with the changing input. It would be easy to focus your attention on the changing neurons and miss the significance of the voting neurons.

The brain wants to reach a consensus. You have probably seen the image above, which can appear as either a vase or two faces. In examples like this, the columns can't determine which is the correct object. It is as if they have two maps for two different towns, but the maps, at least in some areas, are identical. "Vase town" and "Faces town" are similar. The voting layer wants to reach a consensus—it does not permit two objects to be active simultaneously—so it picks one possibility over the other. You can perceive faces or a vase, but not both at the same time.

Attention

It is common that our senses are partially blocked, such as when you look at someone standing behind a car door. Although we only see half a person, we are not fooled. We know that an entire person is standing behind the door. The columns that see the person vote, and they are certain this object is a person. The voting neurons project to the columns whose input is obscured, and now every column knows there is a person. Even the columns that are blocked can predict what they would see if the door wasn't there.

A moment later, we can shift our attention to the car door. Just like the bistable image of the vase and faces, there are two interpretations of the input. We can shift our attention back and forth between "person" and "door." With each shift, the voting neurons settle on a different object. We have the perception that both objects are there, even though we can only attend to one at a time.

The brain can attend to smaller or larger parts of a visual scene. For example, I can attend to the entire car door, or I can attend to just the handle. Exactly how the brain does this is not well understood, but it involves a part of the brain called the thalamus, which is tightly connected to all areas of the neocortex.

Attention plays an essential role in how the brain learns models. As you go about your day, your brain is rapidly and constantly attending to different things. For example, when you read, your attention goes from word to word. Or, looking at a building, your attention can go from building, to window, to door, to door latch, back to door, and so on. What we think is happening is that each time you attend to a different object, your brain determines the object's location relative to the previously attended object. It is automatic. It is part of the attentional process. For example, I enter a dining room. I might first attend to one of the chairs and then to the table. My brain recognizes a chair and then it recognizes a table. However, my brain also calculates the relative position of the chair to the table. As I look around the dining room, my brain is not only recognizing all the objects in the room but simultaneously determining where each object is relative to the other objects and to the room itself. Just by glancing around, my brain builds a model of the room that includes all the objects that I attended to.

Often, the models you learn are temporary. Say you sit down for a family meal in the dining room. You look around the table and see the various dishes. I then ask you to close your eyes and tell me where the potatoes are. You almost certainly will be able to do this, which is proof that you learned a model of the table and its contents in the brief time you looked at it. A few minutes later, after the food has been passed around, I ask you to close your eyes and again point to the potatoes. You will now point to a new location, where you last saw the potatoes. The point of this example is that we are constantly learning models of everything we sense. If the arrangement of features in our models stays fixed, like the logo on the coffee cup, then the model might be remembered for a long time. If the arrangement changes, like the dishes on the table, then the models are temporary.

The neocortex never stops learning models. Every shift of attention—whether you are looking at the dishes on the dining table, walking down the street, or noticing a logo on a coffee cup—is adding another item to a model of something. It is the same learning process if the models are ephemeral or long-lasting.

Hierarchy in the Thousand Brains Theory

For decades, most neuroscientists have adhered to the hierarchy of features theory, and for good reasons. This theory, even though it has many problems, fits a lot of data. Our theory suggests a different way of thinking about the neocortex. The Thousand Brains Theory says that a hierarchy of neocortical regions is not strictly necessary. Even a single cortical region can recognize objects, as evidenced by the mouse's visual system. So, which is it? Is the neocortex organized as a hierarchy or as thousands of models voting to reach a consensus?

The anatomy of the neocortex suggests that both types of connections exist. How can we make sense of this? Our theory suggests a different way of thinking about the connections that is compatible with both hierarchical and single-column models. We have proposed that complete objects, not features, are passed between hierarchical levels. Instead of the neocortex using hierarchy to assemble features into a recognized object, the neocortex uses hierarchy to assemble objects into more complex objects.

I discussed hierarchical composition earlier. Recall the example of a coffee cup with a logo printed on its side. We learn a new object like this by first attending to the cup, then to the logo. The logo is also composed of objects, such as a graphic and a word, but we don't need to remember where the logo's features are relative to the cup. We only need to learn the relative position of the logo's reference frame to the cup's reference frame. All the detailed features of the logo are implicitly included.

This is how the entire world is learned: as a complex hierarchy of objects located relative to other objects. Exactly how the neocortex does this is still unclear. For example, we suspect that some amount of hierarchical learning occurs within each column, but certainly not all of it. Some will be handled by the hierarchical connections between regions. How much is being learned within a single column and how much is being learned in the connections between regions is not understood. We are working on this problem. The answer will almost certainly require a better understanding of attention, which is why we are studying the thalamus.

Earlier in this chapter, I made list of problems with the commonly held view that the neocortex is a hierarchy of feature detectors. Let's go through that list again, this time discussing how the Thousand Brains Theory addresses each problem, starting with the essential role of movement.

- The Thousand Brains Theory is inherently a sensory-motor theory. It explains how we learn and recognize objects by moving. Importantly, it also explains why we can sometimes recognize objects without moving, as when we see a brief image on a screen or grab an object with all our fingers. Thus, the Thousand Brains Theory is a superset of the hierarchical model.
- The relatively large size of regions V1 and V2 in primates and the singularly large size of region V1 in mice makes sense in the Thousand Brains Theory because every column can

recognize complete objects. Contrary to what many neuro-scientists believe today, the Thousand Brains Theory says that most of what we think of as vision occurs in regions V1 and V2. The primary and secondary touch-related regions are also relatively large.

- The Thousand Brains Theory can explain the mystery of how neurons know what their next input will be while the eyes are still in motion. In the theory, each column has models of complete objects and therefore knows what should be sensed at each location on an object. If a column knows the current location of its input and how the eyes are moving, then it can predict the new location and what it will sense there. It is the same as looking at a map of a town and predicting what you will see if you start to walk in a particular direction.

- The binding problem is based on the assumption that the neocortex has a single model for each object in the world. The Thousand Brains Theory flips this around and says that there are thousands of models of every object. The varied inputs to the brain aren't bound or combined into a single model. It doesn't matter that the columns have different types of inputs, or that one column represents a small part of the retina and the next represents a bigger part. It doesn't matter if the retina has holes, any more than it matters that there are gaps between your fingers. The pattern projected to region V1 can be distorted and mixed up and it won't matter, because no part of the neocortex tries to reassemble this scrambled representation. The voting mechanism of the Thousand Brains Theory explains why we have a singular, non-distorted perception. It also explains how recognizing an object in one sensory modality leads to predictions in other sensory modalities.

- Finally, the Thousand Brains Theory shows how the neocor-tex learns three-dimensional models of objects using refer-ence frames. As one more small piece of evidence, look at the following image. It is a bunch of straight lines printed on a flat surface. There is no vanishing point, no converging lines, and no diminishing contrasts to suggest depth. Yet you cannot look at this image without seeing it as a three-dimensional set of stairs. It doesn't matter that the image you are observing is two-dimensional; the models in your neocortex are three-dimensional, and that is what you perceive.

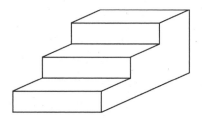

The brain is complex. The details of how place cells and grid cells create reference frames, learn models of environments, and plan behaviors are more complex than I have described, and only partially understood. We are proposing that the neocortex uses similar mechanisms, which are equally complex and even less understood. This is an area of active research for both experimental neuroscientists and theorists like ourselves.

To go further on these and other topics, I would have to introduce additional details of neuroanatomy and neurophysiology, details that are both difficult to describe and not essential for understanding the basics of the Thousand Brains Theory of Intelligence. Therefore, we have reached a border—a border where what this book explores ends, and where what scientific papers need to cover begins.

In the introduction to this book, I said that the brain is like a jigsaw puzzle. We have tens of thousands of facts about the brain, each like a puzzle piece. But without a theoretical framework, we had no idea what the solution to the puzzle looks like. Without a theoretical framework, the best we could do was attach a few pieces together here and there. The Thousand Brains Theory is a framework; it is like finishing the puzzle's border and knowing what the overall picture looks like. As I write, we have filled in some parts of the interior of the puzzle, whereas many other parts are not done. Although a lot remains, our task is simpler now because knowing the proper framework makes it clearer what parts are yet to be filled in.

I don't want to leave you with an incorrect impression that we understand everything the neocortex does. We are far from that. The number of things we don't understand about the brain in general, and the neocortex in particular, is large. However, I don't believe there will be another overall theoretical framework, a different way to arrange the border pieces of the puzzle. Theoretical frameworks get modified and refined over time, and I expect the same will be true for the Thousand Brains Theory, but the core ideas that I presented here will, I believe, mostly remain intact.

————————

Before we leave this chapter and Part 1 of the book, I want to tell you the rest of the story about the time I met Vernon Mountcastle. Recall that I gave a speech at Johns Hopkins University, and at the end of the day I met with Mountcastle and the dean of his department. The time had come for me to leave; I had a flight to catch. We said our goodbyes, and a car was waiting for me outside. As I walked through the office door, Mountcastle intercepted me, put his hand on my shoulder, and said, in a here-is-some-advice-for-you tone of voice, "You should stop talking about hierarchy. It doesn't really exist."

I was stunned. Mountcastle was the world's foremost expert on the neocortex, and he was telling me that one of its largest and most well-documented features didn't exist. I was as surprised as if Francis Crick himself had said to me, "Oh, that DNA molecule, it doesn't really encode your genes." I didn't know how to respond, so I said nothing. As I sat in the car on the way to the airport, I tried to make sense of his parting words.

Today, my understanding of hierarchy in the neocortex has changed dramatically—it is much less hierarchical than I once thought. Did Vernon Mountcastle know this back then? Did he have a theoretical basis for saying that hierarchy didn't really exist? Was he thinking about experimental results that I didn't know about? He died in 2015, and I will never be able to ask him. After his death, I took it upon myself to reread many of his books and papers. His thinking and writing are always insightful. His 1998 *Perceptual Neuroscience: The Cerebral Cortex* is a physically beautiful book and remains one of my favorites about the brain. When I think back on that day, I would have been wise to chance missing my flight for the opportunity to talk to him further. Even more, I wish I could talk to him now. I like to believe he would have enjoyed the theory I just described to you.

Now, I want to turn our attention to how the Thousand Brains Theory will impact our future.

PART 2

Machine Intelligence

In his famous book, *The Structure of Scientific Revolutions*, historian Thomas Kuhn argued that most scientific progress is based on broadly accepted theoretical frameworks, which he called scientific paradigms. Occasionally, an established paradigm is overturned and replaced by a new paradigm—what Kuhn called a scientific revolution.

Today, there are established paradigms for many subfields of neuroscience, such as how the brain evolved, brain-related diseases, and grid cells and place cells. Scientists who work in these fields share terminology and experimental techniques, and they agree on the questions they want to answer. But there is no generally accepted paradigm for the neocortex and intelligence. There is little agreement on what the neocortex does or even what questions we should be trying to answer. Kuhn would say that the study of intelligence and the neocortex is in a pre-paradigm state.

In Part 1 of this book, I introduced a new theory of how the neocortex works and what it means to be intelligent. You could say I am proposing a paradigm for the study of the neocortex. I'm confident that this theory is largely correct, but, importantly, it is also testable. Ongoing and future experiments will tell us which parts of the theory are right and which parts need modification.

In this, the second part of the book, I am going to describe how our new theory will impact the future of artificial intelligence. AI research has an established paradigm, a common set of techniques referred to as artificial neural networks. AI scientists share terminology and goals, which has allowed the field to make steady progress over recent years.

The Thousand Brains Theory of Intelligence suggests that the future of machine intelligence is going to be substantially different from what most AI practitioners are thinking today. I believe that AI is ready for a scientific revolution, and the principles of intelligence that I described earlier will be the basis for that revolution.

I have some hesitation writing about this, due to an experience I had early in my career when I talked about the future of computing. It did not go well.

Shortly after I started Palm Computing, I was asked to give a talk at Intel. Once a year, Intel brought several hundred of its most senior employees to Silicon Valley for a three-day planning meeting. As part of these meetings, they invited a few outside people to address the entire group, and in 1992 I was one of those speakers. I considered it an honor. Intel was leading the personal-computing revolution and was one of the most respected and powerful companies in the world. My company, Palm, was a small start-up that had not yet shipped its first product. My talk was about the future of personal computing.

I proposed that the future of personal computing was going to be dominated by computers small enough to fit in your pocket. These devices would cost somewhere between five hundred and a thousand dollars and run all day on a battery. For billions of people around the world, a pocket-size computer would be the only one they owned. To me, this shift was inevitable. Billions of people wanted access to computers, but laptops and desktops were too expensive and too difficult to use. I saw an inexorable pull toward pocket-size computers, which were easier to use and less expensive.

At that time, there were hundreds of millions of desktop and laptop personal computers. Intel sold the CPUs for most of them. The average CPU chip cost about four hundred dollars and consumed far too much power to be used in a battery-powered, handheld computer. I suggested to the managers at Intel that, if they wanted to continue their leadership position in personal computing, they should focus on three areas: reducing power consumption, making their chips smaller, and figuring out how to make a profit in a product that sold for less than a thousand dollars. The tone of my talk was unassuming, not strident. It was like, "Oh, by the way, I believe this is going to happen and you might want to consider the following implications."

After I finished my talk, I took questions from the audience. Everyone was seated at their lunch table and the food wasn't going to be served until I was finished, so I didn't expect to get many questions. I remember just one. A person stood and asked, in what seemed like a slightly derisive tone, "What are people going to use these handheld computers for?" It was difficult to answer this question.

At that time, personal computers were primarily used for word processing, spreadsheets, and databases. None of these applications were suitable for a handheld computer with a small screen and no keyboard. Logic told me that handheld computers would mostly be used for accessing information, not creating it, and that was the answer I gave. I said that accessing your calendar and address book would be the first applications, but I knew they were not sufficient to transform personal computing. I said we would discover new applications that would be more important.

Recall that in early 1992 there was no digital music, no digital photography, no Wi-Fi, no Bluetooth, and no data on cell phones. The first consumer web browser had yet to be invented. I had no idea that these technologies would be invented, and therefore I could not imagine applications based on them. But I knew that people always wanted more information and that, somehow, we would figure out how to deliver it to mobile computers.

After speaking, I was seated at a table with Dr. Gordon Moore, the legendary founder of Intel. It was a round table with about ten people. I asked Dr. Moore what he thought of my talk. Everyone went silent to hear his response. He avoided giving me a direct answer and then avoided talking to me for the rest of the meal. It soon became clear that neither he nor anyone else at the table believed what I had said.

I was shaken by this experience. If I could not get the smartest and most successful people in computing to even consider my proposal, then perhaps I was wrong, or perhaps the transition to handheld computing would be much harder than I imagined. I decided that the best path forward for me was to focus on building handheld computers, not worrying about what other people believed. Starting that day, I avoided giving "visionary" talks about the future of computing and instead did as much as possible to make that future happen.

Today, I find myself in a similar situation. From here on in the book, I am going to describe a future that is different than what most people, indeed most experts, are expecting. First, I describe a future of artificial intelligence that runs counter to what most of the leaders of AI are currently thinking, and then, in Part 3, I describe the future of humanity in a way you probably have never considered. Of course, I could be wrong; predicting the future is notoriously difficult. But to me, the ideas I am about to present seem inevitable, more like logical deductions than speculation. However, as I experienced at Intel many years ago, I may not be able to convince everyone. I will do my best and ask you to keep an open mind.

In the next four chapters I talk about the future of artificial intelligence. AI is currently undergoing a renaissance. It is one of the hottest fields in technology. Every day seems to bring new applications, new investments, and improved performance. The field of

AI is dominated by artificial neural networks, although they are nothing like the networks of neurons we see in brains. I am going to argue that the future of AI will be based on different principles than those used today, principles that more closely mimic the brain. To build truly intelligent machines, we must design them to adhere to the principles I laid out in the first part of the book.

I don't know what the future applications of AI will be. But like the shift in personal computing to handheld devices, I see the shift of AI toward brain-based principles as inevitable.

CHAPTER 8

Why There Is No "I" in AI

Since its inception in 1956, the field of artificial intelligence has gone through several cycles of enthusiasm followed by pessimism. AI scientists call these "AI summers" and "AI winters." Each wave was based on a new technology that promised to put us on the path to creating intelligent machines, but ultimately these innovations fell short. AI is currently experiencing another wave of enthusiasm, another AI summer, and, once again, expectations in the industry are high. The set of technologies driving the current surge are artificial neural networks, often referred to as deep learning. These methods have achieved impressive results on tasks such as labeling pictures, recognizing spoken language, and driving cars. In 2011, a computer beat the top-ranked humans playing the game show *Jeopardy!*, and in 2016, another computer bested the world's top-ranked player of the game Go. These last two accomplishments made headlines around the world. These achievements are impressive, but are any of these machines truly intelligent?

Most people, including most AI researchers, don't think so. There are numerous ways today's artificial intelligence falls short of human intelligence. For example, humans learn continuously. As I described earlier, we are constantly amending our model of the world. In contrast, deep learning networks have to be fully trained before they can be deployed. And once they are deployed, they can't learn new things on the go. For example, if we want to teach a vision neural network to recognize an additional object, then the network has to be retrained from the ground up, which can take days. However, the biggest reason that today's AI systems are not considered intelligent is they can only do one thing, whereas humans can do many things. In other words, AI systems are not flexible. Any individual human, such as you or me, can learn to play Go, to farm, to write software, to fly a plane, and to play music. We learn thousands of skills in our lifetime, and although we may not be the best at any one of these skills, we are flexible in what we can learn. Deep learning AI systems exhibit almost no flexibility. A Go-playing computer may play the game better than any human, but it can't do anything else. A self-driving car may be a safer driver than any human, but it can't play Go or fix a flat tire.

The long-term goal of AI research is to create machines that exhibit human-like intelligence—machines that can rapidly learn new tasks, see analogies between different tasks, and flexibly solve new problems. This goal is called "artificial general intelligence," or AGI, to distinguish it from today's limited AI. The essential question today's AI industry faces is: Are we currently on a path to creating truly intelligent AGI machines, or will we once again get stuck and enter another AI winter? The current wave of AI has attracted thousands of researchers and billions of dollars of investment. Almost all these people and dollars are being applied to improving deep learning technologies. Will this investment lead to human-level machine intelligence, or are deep learning technologies fundamentally limited, leading us once again to reinvent the field of AI? When you are in the middle of a bubble, it is easy to get swept up in the enthusiasm and believe it will go on forever. History suggests we should be cautious.

I don't know how long the current wave of AI will continue to grow. But I do know that deep learning does not put us on the path to creating truly intelligent machines. We can't get to artificial general intelligence by doing more of what we are currently doing. We have to take a different approach.

Two Paths to AGI

There are two paths that AI researchers have followed to make intelligent machines. One path, the one we are following today, is focused on getting computers to outperform humans on specific tasks, such as playing Go or detecting cancerous cells in medical images. The hope is that if we can get computers to outperform humans on a few difficult tasks, then eventually we will discover how to make computers better than humans at every task. With this approach to AI, it doesn't matter how the system works, and it doesn't matter if the computer is flexible. It only matters that the AI computer performs a specific task better than other AI computers, and ultimately better than the best human. For example, if the best Go-playing computer was ranked sixth in the world, it would not have made headlines and it might even be viewed as a failure. But beating the world's top-ranked human was seen as a major advance.

The second path to creating intelligent machines is to focus on flexibility. With this approach, it isn't necessary that the AI performs better than humans. The goal is to create machines that can do many things and apply what they learn from one task to another. Success along this path could be a machine that has the abilities of a five-year-old child or even a dog. The hope is that if we can first understand how to build flexible AI systems, then, with that foundation, we can eventually make systems that equal or surpass humans.

This second path was favored in some of the earlier waves of AI. However, it proved to be too difficult. Scientists realized that to be as capable as a five-year-old child requires possessing a huge amount of everyday knowledge. Children know thousands of things about the world. They know how liquids spill, balls roll, and dogs bark. They know how to use pencils, markers, paper, and glue. They know how to open books and that paper can rip. They know thousands of words and how to use them to get other people to do things. AI researchers couldn't figure out how to program this everyday knowledge into a computer, or how to get a computer to learn these things.

The difficult part of knowledge is not stating a fact, but representing that fact in a useful way. For example, take the statement "Balls are round." A five-year-old child knows what this means. We can easily enter this statement into a computer, but how can a computer understand it? The words "ball" and "round" have multiple meanings. A ball can be a dance, which isn't round, and a pizza is round, but not like a ball. For a computer to understand "ball," it has to associate the word with different meanings, and each meaning has different relationships to other words. Objects also have actions. For example, some balls bounce, but footballs bounce differently than baseballs, which bounce differently than tennis balls. You and I quickly learn these differences by observation. No one has to tell us how balls bounce; we just throw a ball to the ground and see what happens. We aren't aware of how this knowledge is stored in our brain. Learning everyday knowledge such as how balls bounce is effortless.

AI scientists couldn't figure out how to do this within a computer. They invented software structures called schemas and frames to organize knowledge, but no matter what they tried, they ended up with an unusable mess. The world is complex; the number of things a child knows and the number of links between those things seems impossibly large. I know it sounds like it should be easy, but no one could figure out how a computer could know something as simple as what a ball is.

This problem is called knowledge representation. Some AI scientists concluded that knowledge representation was not only a big problem for AI, it was the *only* problem. They claimed that we could not make truly intelligent machines until we solved how to represent everyday knowledge in a computer.

Today's deep learning networks don't possess knowledge. A Go-playing computer does not know that Go is a game. It doesn't know the history of the game. It doesn't know if it is playing against a computer or a human, or what "computer" and "human" mean. Similarly, a deep learning network that labels images may look at an image and say it is a cat. But the computer has limited knowledge of cats. It doesn't know that cats are animals, or that they have tails, legs, and lungs. It doesn't know about cat people versus dog people, or that cats purr and shed fur. All the deep learning network does is determine that a new image is similar to previously seen images that were labeled "cat." There is no knowledge of cats in the deep learning network.

Recently, AI scientists have tried a different approach to encoding knowledge. They create large artificial neural networks and train them on lots of text: every word in tens of thousands of books, all of Wikipedia, and almost the entire internet. They feed the text into the neural networks one word at a time. By training this way, the networks learn the likelihood that certain words follow other words. These language networks can do some surprising things. For example, if you give the network a few words, it can write a short paragraph related to those words. It is difficult to tell whether the paragraph was written by a human or the neural network.

AI scientists disagree as to whether these language networks possess true knowledge or are just mimicking humans by remembering the statistics of millions of words. I don't believe any kind of deep learning network will achieve the goal of AGI if the network doesn't model the world the way a brain does. Deep learning networks work well, but not because they solved the knowledge representation problem. They work well because they avoided it completely, relying on statistics and lots of data instead. How deep learning networks work is clever, their performance is impressive, and they are commercially valuable. I am only pointing out that they don't possess knowledge and, therefore, are not on the path to having the ability of a five-year-old child.

Brains as a Model for AI

From the moment I became interested in studying the brain, I felt that we would have to understand how it works before we could create intelligent machines. This seemed obvious to me, as the brain is the only thing that we know of that is intelligent. Over the following decades, nothing changed my opinion. That is one reason I have doggedly pursued brain theory: I feel it is a necessary first step to creating truly intelligent AI. I've lived through multiple waves of AI enthusiasm, and each time I resisted jumping on board. It was clear to me that the technologies used were not even remotely like the brain, and therefore AI would get stuck. Figuring out how the brain works is hard, but it is a necessary first step to creating intelligent machines.

In the first half of this book, I described the progress we have made in understanding the brain. I described how the neocortex learns models of the world using maplike reference frames. In the same way that a paper map represents knowledge about a geographic area such as a town or country, the maps in the brain represent knowledge about the objects we interact with (such as bicycles and smartphones), knowledge about our body (such as where our limbs are and how they move), and knowledge about abstract concepts (such as mathematics).

The Thousand Brains Theory solves the problem of knowledge representation. Here is an analogy to help you understand how. Let's say I want to represent knowledge about a common object, a stapler. Early AI researchers would try to do this by listing the names of the different parts of the stapler and then describing what each part does. They might write a rule about staplers that says, "When the top of the stapler is pressed down, a staple comes out of one end." But to understand this statement, words such as "top," "end," and "staple" had to be defined, as did the meaning of the different actions such as "pressed down" and "comes out." And this rule is insufficient on its own. It doesn't say which way the staple faces when it comes out, what happens next, or what you should do if the staple gets stuck. So, the researchers would write additional rules. This method of representing knowledge led to a never-ending list of definitions and rules. AI researchers didn't see how to make it work. Critics argued that even if all the rules could be specified, the computer still wouldn't "know" what a stapler is.

The brain takes a completely different approach to storing knowledge about a stapler: it learns a model. The model is the embodiment of knowledge. Imagine for a moment that there is a tiny stapler in your head. It is exactly like a real stapler—it has the same shape, the same parts, and it moves in the same ways— it's just smaller. The tiny model represents everything you know about staplers without needing to put a label on any of the parts. If you want to recall what happens when the top of a stapler is pressed down, you press down on the miniature model and see what happens.

Of course, there isn't a tiny physical stapler in your head. But the cells in the neocortex learn a virtual model that serves the same purpose. As you interact with a real stapler, the brain learns its virtual model, which includes everything you have observed about the real stapler, from its shape to how it behaves when you use it. Your knowledge of staplers is embedded in the model. There isn't a list of stapler facts and stapler rules stored in your brain.

Let's say I ask you what happens when the top of a stapler is pushed down. To answer this question, you don't find the appropriate rule and play it back to me. Instead, your brain imagines pressing down on the stapler, and the model recalls what happens. You can use words to describe it to me, but the knowledge is not stored in words or rules. The knowledge is the model.

I believe the future of AI will be based on brain principles. Truly intelligent machines, AGI, will learn models of the world using maplike reference frames just like the neocortex. I see this as inevitable. I don't believe there is another way to create truly intelligent machines.

Moving from Dedicated to Universal AI Solutions

The situation we are in today reminds me of the early days of computing. The word "computer" originally referred to people whose job was to perform mathematical calculations. To create numeric tables or to decode encrypted messages, dozens of human computers would do the necessary calculations by hand. The very first electronic computers were designed to replace human computers for a specific task. For example, the best automated solution for message decryption was a machine that only decrypted messages. Computing pioneers such as Alan Turing argued that we should build "universal" computers: electronic machines that could be programmed to do any task. However, at that time, no one knew the best way to build such a computer.

There was a transitionary period where computers were built in many different forms. There were computers designed for specific tasks. There were analog computers, and computers that could only be repurposed by changing the wiring. There were computers that worked with decimal instead of binary numbers. Today, almost all computers are the universal form that Turing envisioned. We even refer to them as "universal Turing machines." With the right software, today's computers can be applied to almost any task. Market forces decided that universal, general-purpose computers were the way to go. This is despite the fact that, even today, any particular task can be performed faster or with less power using a custom solution, such as a special chip. Product designers and engineers usually prefer the lower cost and convenience of general-purpose computers, even though a dedicated machine could be faster and use less power.

A similar transition will occur with artificial intelligence. Today we are building dedicated AI systems that are the best at whatever task they are designed to do. But in the future, most intelligent machines will be universal: more like humans, capable of learning practically anything.

Today's computers come in many shapes and sizes, from the microcomputer in a toaster to room-size computers used for weather simulation. Despite their differences in size and speed, all these computers work on the same principles laid out by Turing and others many years ago. They are all instances of universal Turing machines. Similarly, intelligent machines of the future will come in many shapes and sizes, but almost all of them will work on a common set of principles. Most AI will be universal learning machines, similar to the brain. (Mathematicians have proven that there are some problems that cannot be solved, even in principle. Therefore, to be precise, there are no true "universal" solutions. But this is a highly theoretical idea and we don't need to consider it for the purposes of this book.)

Some AI researchers argue that today's artificial neural networks are already universal. A neural network can be trained to play Go or drive a car. However, the same neural network can't do both. Neural networks also have to be tweaked and modified in other ways to get them to perform a task. When I use the terms "universal" or "general-purpose," I imagine something like ourselves: a machine that can learn to do many things without erasing its memory and starting over.

There are two reasons AI will transition from the dedicated solutions we see today to more universal solutions that will dominate the future. The first is the same reason that universal computers won out over dedicated computers. Universal computers are ultimately more cost-effective, and this led to more rapid advances in the technology. As more and more people use the same designs, more effort is applied to enhancing the most popular designs and the ecosystems that support them, leading to rapid improvements in cost and performance. This was the underlying driver of the exponential increase in computing power that shaped industry and society in the latter part of the twentieth century. The second reason that AI will transition to universal solutions is that some of the most important future applications of machine intelligence will require the flexibility of universal solutions. These applications will need to handle unanticipated problems and devise novel solutions in a way that today's dedicated deep learning machines cannot.

Consider two types of robots. The first robot paints cars in a factory. We want car-painting robots to be fast, accurate, and unchanging. We don't want them trying new spraying techniques each day or questioning why they are painting cars. When it comes to painting cars on an assembly line, single-purpose, unintelligent robots are what we need. Now say we want to send a team of robot construction workers to Mars to build a livable habitat for humans. These robots need to use a variety of tools and assemble buildings in an unstructured environment. They will encounter unforeseen problems and will need to collaboratively improvise fixes and modify designs. Humans can handle these types of issues, but no machine today is close to doing any of this. Mars construction robots will need to possess general-purpose intelligence.

You might think that the need for general-purpose intelligent machines will be limited, that most AI applications will be addressed with dedicated, single-purpose technologies like we have today. People thought the same thing about general-purpose computers. They argued that the commercial demand for general-purpose computers was limited to a few high-value applications. The opposite turned out to be true. Due to dramatic reductions in cost and size, general-purpose computers became one of the largest and most economically important technologies of the last century. I believe that general-purpose AI will similarly dominate machine intelligence in the latter part of the twenty-first century. In the late 1940s and early 1950s, when commercial computers first became available, it was impossible to imagine what their applications would be in 1990 or 2000. Today, our imagination is similarly challenged. No one can know how intelligent machines will be used fifty or sixty years from now.

When Is Something Intelligent?

When should we consider a machine intelligent? Is there a set of criteria we can use? This is analogous to asking, When is a machine a general-purpose computer? To qualify as a general-purpose computer—that is, a universal Turing machine—a machine needs certain components, such as memory, a CPU, and software. You can't detect these ingredients from the outside. For example, I can't tell if my toaster oven has a general-purpose computer inside or a custom chip. The more features my toaster oven has, the more likely it contains a general-purpose computer, but the only sure way to tell is by looking inside and seeing how it works.

Similarly, to qualify as intelligent, a machine needs to operate using a set of principles. You can't detect whether a system uses these principles by observing it from the outside. For example, if I see a car driving down the highway, I can't tell if it is being driven by an intelligent human who is learning and adapting as they drive or by a simple controller that just keeps the car between two lines. The more complex the behavior exhibited by the car, the more likely it is that an intelligent agent is in control, but the only sure way to tell is by looking inside.

So, is there a set of criteria that machines must have to be considered intelligent? I think so. My proposal for what qualifies as intelligent is based on the brain. Each of the four attributes in the following list is something we know that the brain does and that I believe an intelligent machine must do too. I will describe what each attribute is, why it is important, and how the brain implements it. Of course, intelligent machines may implement these attributes differently than a brain. For example, intelligent machines don't have to be made of living cells.

Not everyone will agree with my choice of attributes. One can make a good argument that I have left some important things out. That's OK. I view my list as a minimum, or baseline, for AGI. Few AI systems have any of these attributes today.

1. Learning Continuously

What is it? Every waking moment of our entire life, we are learning. How long we remember something varies. Some things are forgotten quickly, such as the arrangement of dishes on a table or what clothes we wore yesterday. Other things will stay with us for our entire lives. Learning is not a separate process from sensing and acting. We learn continuously.

Why is it important? The world is constantly changing; therefore, our model of the world must learn continuously to reflect the changing world. Most AI systems today do not learn continuously. They go through a lengthy training process and when it is complete, they are deployed. This is one reason they are not flexible. Flexibility requires continuously adjusting to changing conditions and new knowledge.

How does the brain do it? The most important component of how brains learn continuously is the neuron. When a neuron learns a new pattern, it forms new synapses on one dendrite branch. The new synapses don't affect previously learned ones on other branches. Thus, learning something new doesn't force the neuron to forget or modify something it learned earlier. The artificial neurons used in today's AI systems don't have this ability. This is one reason they can't learn continuously.

2. Learning via Movement

What is it? We learn by moving. As we go about our day, we move our body, our limbs, and our eyes. These movements are integral to how we learn.

Why is it important? Intelligence requires learning a model of the world. We cannot sense everything in the world at once; therefore, movement is required for learning. You cannot learn a model of a house without moving from room to room and you cannot learn a new app on your smartphone without interacting with it. Movements don't have to be physical. The principle of learning via movement also applies to concepts such as mathematics and to virtual spaces such as the internet.

How does the brain do it? The unit of processing in the neocortex is the cortical column. Each column is a complete sensory-motor system—that is, it gets inputs and it can generate behaviors. With every movement, a column predicts what its next input will be. Prediction is how a column tests and updates its model.

3. Many Models

What is it? The neocortex is composed of tens of thousands of cortical columns, and each column learns models of objects. Knowledge about any particular thing, such as a coffee cup, is distributed among many complementary models.

Why is it important? The many-models design of the neocortex provides flexibility. By adopting this architecture, AI designers can easily create machines that integrate multiple types of sensors—such as vision and touch, or even novel sensors such as radar. And they can create machines that have varied embodiments. Like the neocortex, the "brain" of an intelligent machine will consist of many nearly identical elements that can then be connected to a variety of moveable sensors.

How does the brain do it? The key to making the many-models design work is voting. Each column operates somewhat independently, but long-range connections in the neocortex allow columns to vote on what object they are sensing.

4. Using Reference Frames to Store Knowledge

What is it? In the brain, knowledge is stored in reference frames. Reference frames are also used to make predictions, create plans, and perform movements. Thinking occurs as the brain activates one location at a time in a reference frame and the associated piece of knowledge is retrieved.

Why is it important? To be intelligent, a machine needs to learn a model of the world. That model must include the shape of objects, how they change as we interact with them, and where they are relative to each other. Reference frames are needed to represent this kind of information; they are the backbone of knowledge.

How does the brain do it? Each cortical column establishes its own set of reference frames. We have proposed that cortical columns create reference frames using cells that are equivalent to grid cells and place cells.

Examples of Reference Frames

Most artificial neural networks have nothing equivalent to reference frames. For example, a typical neural network that recognizes images just assigns a label to each image. Without reference frames, the network has no way of learning the 3D structure of objects or how they move and change. One of the problems with a system like this is that we can't ask why it labeled something a cat. The AI system doesn't know what a cat is. There is no further information to be had, beyond that this image is similar to other images that were labeled "cat."

Some forms of AI do have reference frames, although the way they are implemented is limiting. For example, a computer that plays chess has a reference frame: the chessboard. Locations on a chessboard are noted in chess-specific nomenclature, such as "king's rook 4" or "queen 7." The chess-playing computer uses this reference frame to represent the location of each piece, to represent legal chess moves, and to plan movements. A chessboard reference frame is inherently two-dimensional and has only sixty-four locations. This is fine for chess, but it is useless for learning the structure of staplers or the behaviors of cats.

Self-driving cars typically have multiple reference frames. One is GPS, the satellite-based system that can locate the car anywhere on Earth. Using a GPS reference frame, a car can learn where roads, intersections, and buildings are. GPS is a more general-purpose reference frame than a chess board, but it is anchored to Earth, and therefore can't represent the structure or shape of things that move relative to Earth, such as a kite or a bicycle.

Robot designers are accustomed to using reference frames. They use them to keep track of where a robot is in the world and to plan how it should move from one location to another. Most roboticists are not concerned about AGI, whereas most AI researchers are unaware of the importance of reference frames. Today, AI and robotics are largely separate fields of research, although the line is starting to blur. Once AI researchers understand the essential role of movement and reference frames for creating AGI, the separation between artificial intelligence and robotics will disappear completely.

One AI scientist who understands the importance of reference frames is Geoffrey Hinton. Today's neural networks rely on ideas that Hinton developed in the 1980s. Recently, he has become critical of the field because deep learning networks lack any sense of location and, therefore, he argues, they can't learn the structure of the world. In essence, this is the same criticism I am making, that AI needs reference frames. Hinton has proposed a solution to this problem that he calls "capsules." Capsules promise dramatic improvements in neural networks, but so far they have not caught on in mainstream applications of AI. Whether capsules succeed or whether future AI relies on grid-cell-like mechanisms as I have proposed remains to be seen. Either way, intelligence requires reference frames.

Finally, let's consider animals. All mammals have a neocortex and therefore are all, by my definition, intelligent, general-purpose learners. Every neocortex, big or small, has general-purpose reference frames defined by cortical grid cells.

A mouse has a small neocortex. Therefore, the capacity of what it can learn is limited compared to an animal with a larger neocortex. But I would say a mouse is intelligent in the same way that the computer in my toaster is a universal Turing machine. The toaster's computer is a small, yet complete, implementation of Turing's idea. Similarly, a mouse brain is a small, yet complete, implementation of the learning attributes described in this chapter.

Intelligence in the animal world is not limited to mammals. For example, birds and octopuses learn and exhibit complex behaviors. It is almost certain that these animals also have reference frames in their brains, although it remains to be discovered whether they have something like grid cells and place cells or a different mechanism.

These examples show that almost every system that exhibits planning and complex, goal-oriented behavior—whether it be a chess-playing computer, a self-driving car, or a human—has reference frames. The type of reference frame dictates what the system can learn. A reference frame designed for a specific task, such as playing chess, will not be useful in other domains. General-purpose intelligence requires general-purpose reference frames that can be applied to many types of problems.

It is worth emphasizing again that intelligence cannot be measured by how well a machine performs a single task, or even several tasks. Instead, intelligence is determined by how a machine learns and stores knowledge about the world. We are intelligent not because we can do one thing particularly well, but because we can learn to do practically anything. The extreme flexibility of human intelligence requires the attributes I described in this chapter: continuous learning, learning through movement, learning many models, and using general-purpose reference frames for storing knowledge and generating goal-oriented behaviors. In the future, I believe almost all forms of machine intelligence will have these attributes, although we are a long way from that today.

There is a group of people who will argue that I have ignored the most important topic related to intelligence: consciousness. I will address this in the next chapter.

CHAPTER 9

When Machines Are Conscious

I recently attended a panel discussion titled "Being Human in the Age of Intelligent Machines." At one point during the evening, a philosophy professor from Yale said that if a machine ever became conscious, then we would probably be morally obligated to not turn it off. The implication was that if something is conscious, even a machine, then it has moral rights, so turning it off is equivalent to murder. Wow! Imagine being sent to prison for unplugging a computer. Should we be concerned about this?

Most neuroscientists don't talk much about consciousness. They assume that the brain can be understood like every other physical system, and consciousness, whatever it is, will be explained in the same way. Since there isn't even an agreement on what the word "consciousness" means, it is best to not worry about it.

Philosophers, on the other hand, love to talk (and write books) about consciousness. Some believe that consciousness is beyond physical description. That is, even if you had a full understanding of how the brain works, it would not explain consciousness. Philosopher David Chalmers famously claimed that consciousness is "the hard problem," whereas understanding how the brain works is "the easy problem." This phrase caught on, and now many people just assume that consciousness is an inherently unsolvable problem.

Personally, I see no reason to believe that consciousness is beyond explanation. I don't want to get into debates with philosophers, nor do I want to try to define consciousness. However, the Thousand Brains Theory suggests physical explanations for several aspects of consciousness. For example, the way the brain learns models of the world is intimately tied to our sense of self and how we form beliefs.

What I want to do in this chapter is describe what brain theory says about a few aspects of consciousness. I will stick to what we know about the brain and let you decide what, if anything, remains to be explained.

Awareness

Imagine if I could reset your brain to the exact state it was in when you woke up this morning. Before I reset you, you would get up and go about your day, doing the things you normally do. Perhaps on this day you washed your car. At dinnertime, I would reset your brain to the time you got up, undoing any changes—including any changes to the synapses—that occurred during the day. Therefore, all memories of what you did would be erased. After I reset your brain, you would believe that you just woke up. If I then told you that you had washed your car today, you would at first protest, claiming it wasn't true. Upon showing you a video of you washing your car, you might admit that it indeed looks like you had, but you could not have been conscious at the time. You might also claim that you shouldn't be held responsible for anything you did during the day because you were not conscious when you did it. Of course, you *were* conscious when you washed your car. It is only after deleting your memories of the day that you would believe and claim you were not. This thought experiment shows that our sense of awareness, what many people would call being conscious, requires that we form moment-to-moment memories of our actions.

Consciousness also requires that we form moment-to-moment memories of our thoughts. Recall that thinking is just a sequential activation of neurons in the brain. We can remember a sequence of thoughts just as we can remember the sequence of notes in a melody. If we didn't remember our thoughts, we would be unaware of why we were doing anything. For example, we have all experienced going to a room in our house to do something but, upon entering the room, forgetting what we went there for. When this happens, we often ask ourselves, "Where was I just before I got here and what was I thinking?" We try to recall the memory of our recent thoughts so we know why we are now standing in the kitchen.

When our brains are working properly, the neurons form a continuous memory of both our thoughts and actions. Therefore, when we get to the kitchen, we can recall the thoughts we had earlier. We retrieve the recently stored memory of thinking about eating the last piece of cake in the refrigerator and we know why we went to the kitchen.

The active neurons in the brain at some moments represent our present experience, and at other moments represent a previous experience or a previous thought. It is this accessibility of the past—the ability to jump back in time and slide forward again to the present—that gives us our sense of presence and awareness. If we couldn't replay our recent thoughts and experiences, then we would be unaware that we are alive.

Our moment-to-moment memories are not permanent. We typically forget them within hours or days. I remember what I had for breakfast today, but I will lose this memory in a day or two. It is common that our ability to form short-term memories declines with age. That is why we have more and more of the "Why did I come here?" experiences as we get older.

These thought experiments prove that our awareness, our sense of presence—which is the central part of consciousness—is dependent on continuously forming memories of our recent thoughts and experiences and playing them back as we go about our day.

Now let's say we create an intelligent machine. The machine learns a model of the world using the same principles as a brain. The internal states of the machine's model of the world are equivalent to the states of neurons in the brain. If our machine remembers these states as they occur and can replay these memories, then would it be aware and conscious of its existence, in the same way that you and I are? I believe so.

If you believe that consciousness cannot be explained by scientific investigation and the known laws of physics, then you might argue that I have shown that storing and recalling the states of a brain is necessary, but I have not proven that is sufficient. If you take this view, then the burden is on you to show why it is not sufficient.

For me, the sense of awareness—the sense of presence, the feeling that I am an acting agent in the world—is the core of what it means to be conscious. It is easily explained by the activity of neurons, and I see no mystery in it.

Qualia

The nerve fibers that enter the brain from the eyes, ears, and skin look the same. Not only do they look identical, they transmit information using identical-looking spikes. If you look at the inputs to the brain, you can't discern what they represent. Yet, vision feels like one thing and hearing feels like something different, and neither feels like spikes. When you look at a pastoral scene you don't sense the *tat-tat-tat* of electrical spikes entering your brain; you see hills and color and shadows.

"Qualia" is the name for how sensory inputs are perceived, how they feel. Qualia are puzzling. Given that all sensations are created by identical spikes, why does seeing feel different than touching? And why do some input spikes result in the sensation of pain and others don't? These may seem like silly questions, but if you imagine that the brain is sitting in the skull and its inputs are just spikes, then you can get a sense of the mystery. Where do our perceived sensations come from? The origin of qualia is considered one of the mysteries of consciousness.

Qualia Are Part of the Brain's Model of the World

Qualia are subjective, which means they are internal experiences. For example, I know what a pickle tastes like to me, but I can't know if a pickle tastes the same to you. Even if we use the same words to describe the taste of a pickle, it is still possible that you and I perceive pickles differently. Sometimes we actually know that the same input is perceived differently by different people. A recent famous example of this is a photograph of a dress that some people see as white and gold and other people see as black and blue. The exact same picture can result in different perceptions of color. This tells us that the qualia of color is not purely a property of the physical world. If it were, we would all say the dress has the same color. The color of the dress is a property of our brain's model of the world. If two people perceive the same input differently, that tells us their model is different.

There is a fire station near my home that has a red fire truck outside in the driveway. The surface of the truck always appears red, even though the frequency and intensity of the light reflected off it varies. The light changes with the angle of the Sun, the weather, and the orientation of the truck in the driveway. Yet I don't perceive the color of the truck changing. This tells us that there isn't a one-to-one correspondence between what we perceive as red and a particular frequency of light. Red is *related* to particular frequencies of light, but what we perceive as red is not always the same frequency. The redness of the fire truck is a fabrication of the brain—it is a property of the brain's model of surfaces, not a property of light per se.

Some Qualia Are Learned via Movements, Similar to How We Learn Objects

If qualia are a property of the brain's model of the world, then how does the brain create them? Recall that the brain learns models of the world by movement. To learn what a coffee cup feels like, you must move your fingers over the coffee cup, touching it at different locations.

Some qualia are learned in a similar way, via movement. Imagine you are holding a green piece of paper in your hand. As you look at it, you move it. First you look at the paper straight on, then you angle it to the left, then right, then up, then down. As you change the angle of the paper, the frequency and intensity of light entering your eyes changes, and thus the pattern of spikes entering your brain also changes. As you move an object, such as the piece of green paper, your brain predicts how the light will change. We can be certain this prediction is occurring, because if the light didn't change or it changed differently than normal as you moved the paper, you would notice that something was wrong. This is because the brain has models for how surfaces reflect light at different angles. There are different models for different types of surfaces. We might call the model of one surface "green" and another "red."

How would a model of a surface's color be learned? Imagine we have a reference frame for surfaces we call green. The reference frame for green is different than the reference frame for an object, such as a coffee cup, in one important way: A reference frame for the cup represents sensed inputs at different *locations* on the cup. A reference frame for a green surface represents sensed inputs at different *orientations* of the surface. You might find it difficult to imagine a reference frame that represents orientations, but from a theory point of view, the two types of reference frames are similar. The same basic mechanism that the brain uses to learn models of coffee cups could also learn models of colors.

Without further evidence, I don't know if the qualia of color are actually modeled this way. I mention this example because it shows that it is possible to construct testable theories and neural explanations for how we learn and experience qualia. It shows that qualia may not lie outside of the realm of normal scientific explanation, as some people believe.

Not all qualia are learned. For example, the feeling of pain is almost certainly innate, mediated by special pain receptors and old-brain structures, not the neocortex. If you touch a hot stove, your arm will retract in pain before your neocortex knows what is happening. Therefore, pain can't be understood in the same way as the color green, which I am proposing is learned in the neocortex.

When we feel pain, it is "out there," at some location on our body. Location is part of the qualia of pain, and we have a solid explanation for why it is perceived at different locations. But I don't have an explanation for why pain hurts, or why it feels the way it does and not like something else. This doesn't bother me in any deep way. There are many things we don't yet understand about the brain, but the steady progress we have made gives me confidence that these and other issues related to qualia can be understood in the normal course of neuroscience research and discovery.

The Neuroscience of Consciousness

There are neuroscientists who study consciousness. On one end of a spectrum are neuroscientists who believe that consciousness is likely to be beyond normal scientific explanation. They study the brain to look for neural activity that *correlates* with consciousness, but they do not believe neural activity can *explain* it. They suggest that perhaps consciousness can never be understood, or maybe it is created by quantum effects or undiscovered laws of physics. Personally, I can't understand this point of view. Why would we ever assume something can't be understood? The long history of human discovery has shown again and again that things that at first appear to be beyond comprehension ultimately have logical explanations. If a scientist makes the extraordinary claim that consciousness cannot be explained by neural activity, we should be skeptical, and the burden should be on them to show why.

There are other neuroscientists who study consciousness who believe that it can be understood like any other physical phenomenon. If consciousness seems mysterious, they argue, it is only because we don't yet understand the mechanisms, and perhaps we are not thinking about the problem correctly. My colleagues and I are squarely in this camp. So is Princeton neuroscientist Michael Graziano. He has proposed that a particular region of the neocortex models attention, similar to how somatic regions of the neocortex model the body. He proposes that the brain's model of attention leads us to *believe* we are conscious, in the same way that the brain's model of the body leads us to *believe* we have an arm or a leg. I don't know if Graziano's theory is correct, but to me, it represents the right approach. Notice that his theory is based on the neocortex learning a *model* of attention. If he is right, I would wager that that model is built using grid-cell-like reference frames.

Machine Consciousness

If it is true that consciousness is just a physical phenomenon, then what should we expect about intelligent machines and consciousness? I have no doubts that machines that work on the same principles as the brain will be conscious. AI systems don't work this way today, but in the future they will, and they will be conscious. I also have no doubts that many animals, especially other mammals, are also conscious. They don't have to tell us for us to know; we can tell they are conscious by seeing that their brains work similarly to ours.

Do we have a moral obligation to not turn off a conscious machine? Would that be equivalent to murder? No. I would have no concerns about unplugging a conscious machine. First, consider that we humans turn off every night when we go to sleep. We turn on again when we wake. That, in my mind, is no different than unplugging a conscious machine and plugging it in again later.

What about destroying an intelligent machine when it is unplugged, or just never plugging it in again? Wouldn't that be akin to murdering a person while they are asleep? Not really.

Our fear of death is created by the older parts of our brain. If we detect a life-threatening situation, then the old brain creates the sensation of fear and we start acting in more reflexive ways. When we lose someone close to us, we mourn and feel sad. Fears and emotions are created by neurons in the old brain when they release hormones and other chemicals into the body. The neocortex may help the old brain decide when to release these chemicals, but without the old brain we would not sense fear or sadness. Fear of death and sorrow for loss are not required ingredients for a machine to be conscious or intelligent. Unless we go out of our way to give machines equivalent fears and emotions, they will not care at all if they are shut down, disassembled, or scrapped.

It is possible that a human could become attached to an intelligent machine. Perhaps they shared many experiences, and the human feels a personal connection with it. In that case, we would have to consider the harm caused to the human if we turned off the machine. But there wouldn't be a moral obligation to the intelligent machine itself. If we went out of our way to give intelligent machines fears and emotions, then I would take a different position, but intelligence and consciousness on their own do not create this kind of moral dilemma.

The Mystery of Life and the Mystery of Consciousness

Not that long ago, the question "What is life?" was as mysterious as "What is consciousness?" It seemed impossible to explain why some pieces of matter were alive and others were not. To many people, this mystery seemed beyond scientific explanation. In 1907, philosopher Henri Bergson introduced a mysterious thing he called *élan vital* to explain the difference between living and nonliving things. According to Bergson, inanimate matter became living matter with the addition of *élan vital*. Importantly, *élan vital* was not physical and could not be understood by normal scientific study.

With the discovery of genes, DNA, and the entire field of biochemistry, we no longer view living matter as unexplainable. There are still many unanswered questions about life, such as how did it begin, is it common in the universe, is a virus a living thing, and can life exist using different molecules and chemistry? But these questions, and the debates they create, are on the edge. Scientists no longer debate whether life is explainable. At some point it became clear that life can be understood as biology and chemistry. Concepts like *élan vital* became a part of history.

I expect that a similar change of attitude will occur with consciousness. At some point in the future, we will accept that any system that learns a model of the world, continuously remembers the states of that model, and recalls the remembered states will be conscious. There will be remaining unanswered questions, but consciousness will no longer be talked about as "the hard problem." It won't even be considered a problem.

CHAPTER 10

The Future of
Machine Intelligence

Nothing we call AI today is intelligent. No machine exhibits the flexible modeling capabilities I described in the earlier chapters of this book. Yet there are no technical reasons preventing us from creating intelligent machines. The obstacles have been a lack of understanding of what intelligence is, and not knowing the mechanisms needed to create it. By studying how the brain works, we have made significant progress addressing these issues. It seems inevitable to me that we will overcome any remaining obstacles and enter the age of machine intelligence in this century, probably in the next two to three decades.

Machine intelligence will transform our lives and our society. I believe it will have a larger impact on the twenty-first century than computing did on the twentieth. But, as with most new technologies, it is impossible to know exactly how this transformation will play out. History suggests that we can't anticipate the technological advances that will push machine intelligence forward. Think of the innovations that drove the acceleration of computing, such as the integrated circuit, solid state memory, cellular wireless communications, public-key cryptography, and the internet. Nobody in 1950 anticipated these and many other advances. Similarly, nobody anticipated how computers would transform media, communications, and commerce. I believe we are similarly ignorant today about what intelligent machines will look like and how we will use them seventy years from now.

Although we can't know the details of the future, the Thousand Brains Theory can help us define the boundaries. Understanding how the brain creates intelligence tells us what things are possible, what things are not, and to some extent what advances are likely. That is the goal of this chapter.

Intelligent Machines Will Not Be Like Humans

The most important thing to keep in mind when thinking about machine intelligence is the major division of the brain that I discussed in Chapter 2: the old brain versus the new brain. Recall that the older parts of the human brain control the basic functions of life. They create our emotions, our desires to survive and procreate, and our innate behaviors. When creating intelligent machines, there is no reason we should replicate all the functions of the human brain. The new brain, the neocortex, is the organ of intelligence, so intelligent machines need something equivalent to it. When it comes to the rest of the brain, we can choose which parts we want and which parts we don't.

Intelligence is the ability of a system to learn a model of the world. However, the resulting model by itself is valueless, emotionless, and has no goals. Goals and values are provided by whatever system is using the model. It's similar to how the explorers of the sixteenth through the twentieth centuries worked to create an accurate map of Earth. A ruthless military general might use the map to plan the best way to surround and murder an opposing army. A trader could use the exact same map to peacefully exchange goods. The map itself does not dictate these uses, nor does it impart any value to how it is used. It is just a map, neither murderous nor peaceful. Of course, maps vary in detail and in what they cover. Therefore, some maps might be better for war and others better for trade. But the desire to wage war or trade comes from the person using the map.

Similarly, the neocortex learns a model of the world, which by itself has no goals or values. The emotions that direct our behaviors are determined by the old brain. If one human's old brain is aggressive, then it will use the model in the neocortex to better execute aggressive behavior. If another person's old brain is benevolent, then it will use the model in the neocortex to better achieve its benevolent goals. As with maps, one person's model of the world might be better suited for a particular set of aims, but the neocortex does not create the goals.

Intelligent machines need to have a model of the world and the flexibility of behavior that comes from that model, but they don't need to have human-like instincts for survival and procreation. In fact, designing a machine to have human-like emotions is far more difficult than designing one to be intelligent, because the old brain comprises numerous organs, such as the amygdala and hypothalamus, each of which has its own design and function. To make a machine with human-like emotions, we would have to recreate the varied parts of the old brain. The neocortex, although much larger than the old brain, comprises many copies of a relatively small element, the cortical column. Once we know how to build one cortical column, it should be relatively easy to put lots of them into a machine to make it more intelligent.

The recipe for designing an intelligent machine can be broken into three parts: embodiment, parts of the old brain, and the neocortex. There is a lot of latitude in each of these components, and therefore there will be many types of intelligent machines.

1. Embodiment

As I described earlier, we learn by moving. In order to learn a model of a building, we must walk through it, going from room to room. To learn a new tool, we must hold it in our hand, turning it this way and that, looking and attending to different parts with our fingers and eyes. At a basic level, to learn a model of the world requires moving one or more sensors relative to the things in the world.

Intelligent machines also need sensors and the ability to move them. This is referred to as embodiment. The embodiment could be a robot that looks like a human, a dog, or a snake. The embodiment could take on nonbiological forms, such as a car or a ten-armed factory robot. The embodiment can even be virtual, such as a bot exploring the internet. The idea of a virtual body may sound strange. The requirement is that an intelligent system can perform actions that change the locations of its sensors, but actions and locations don't have to be physical. When you browse on the Web you move from one location to another, and what you sense changes with each new website. We do this by physically moving a mouse or touching a screen, but an intelligent machine could do the same just using software, with no physical movements. Most of today's deep learning networks don't have an embodiment. They don't have moveable sensors and they don't have reference frames to know where the sensors are. Without embodiment, what can be learned is limited.

The types of sensors that can be used in an intelligent machine are almost limitless. A human's primary senses are vision, touch, and hearing. Bats have sonar. Some fish have senses that emit electric fields. Within vision, there are eyes with lenses (like ours), compound eyes, and eyes that see infrared or ultraviolet light. It is easy to imagine new types of sensors designed for specific problems. For example, a robot capable of rescuing people in collapsed buildings might have radar sensors so it can see in the dark.

Human vision, touch, and hearing are achieved through arrays of sensors. For example, an eye is not a single sensor. It contains thousands of sensors arrayed on the back of the eye. Similarly, the body contains thousands of sensors arrayed on the skin. Intelligent machines will also have sensory arrays. Imagine if you only had one finger for touching, or you could only look at the world through a narrow straw. You would still be able to learn about the world, but it would take much longer, and the actions you could perform would be limited. I can imagine simple intelligent machines with limited capabilities having just a few sensors, but a machine that approaches or exceeds human intelligence will have large sensory arrays—just as we do.

Smell and taste are qualitatively different than vision and touch. Unless we put our nose directly on a surface, as dogs do, it is difficult to say where a smell is located with any precision. Similarly, taste is limited to sensing things in the mouth. Smell and taste help us decide what foods are safe to eat, and smell might help us identify a general area, but we don't rely on them much for learning the detailed structure of the world. This is because we can't easily associate smells and tastes with specific locations. This is not an inherent limitation of these senses. For example, an intelligent machine could have arrays of taste-like chemical sensors on the surface of its body, allowing the machine to "feel" chemicals in the same way you and I feel textures.

Sound is in between. By using two ears and taking advantage of how sound bounces off our outer ear, our brains can locate sounds much better than they can locate smell or taste, but not as well as with vision and touch.

The important point is that for an intelligent machine to learn a model of the world, it needs sensory inputs that can be moved. Each individual sensor needs to be associated with a reference frame that tracks the location of the sensor relative to things in the world. There are many different types of sensors that an intelligent machine could possess. The best sensors for any particular application depend on what kind of world the machine exists in and what we hope the machine will learn.

In the future, we might build machines with unusual embodiments. For example, imagine an intelligent machine that exists inside individual cells and understands proteins. Proteins are long molecules that naturally fold into complex shapes. The shape of a protein molecule determines what it does. There would be tremendous benefits to medicine if we could better understand the shape of proteins and manipulate them as needed, but our brains are not very good at understanding proteins. We cannot sense them or interact with them directly. Even the speed at which they act is much faster than our brains can process. However, it might be possible to create an intelligent machine that understands and manipulates proteins in the same way that you and I understand and manipulate coffee cups and smartphones. The brain of the intelligent protein machine (IPM) might reside in a typical computer, but its movements and sensors would work on a very small scale, inside a cell. Its sensors might detect amino acids, different types of protein folds, or particular chemical bonds. Its actions might involve moving its sensors relative to a protein, as you might move your finger over a coffee cup. And it might have actions that prod a protein to get it to change its shape, similar to how you touch a smartphone screen to change its display. The IPM could learn a model of the world inside of cells and then use this model to achieve desired goals, such as eliminating bad proteins and fixing damaged ones.

Another example of an unusual embodiment is a distributed brain. The human neocortex has about 150,000 cortical columns, each modeling the part of the world that it can sense. There is no reason that the "columns" of an intelligent machine must be physically located next to each other, as in a biological brain. Imagine an intelligent machine with millions of columns and thousands of sensor arrays. The sensors and the associated models could be physically distributed across the Earth, within the oceans, or throughout our solar system. For example, an intelligent machine with sensors distributed over the surface of Earth might understand the behavior of global weather in the same way you and I understand the behavior of a smartphone.

I don't know whether it will ever be feasible to build an intelligent protein machine or how valuable distributed intelligent machines will be. I mention these examples to stimulate your imagination and because they are in the realm of possibility. The key idea is that intelligent machines will likely take many different forms. When we think about the future of machine intelligence and the implications it will have, we need to think broadly and not limit our ideas to the human and other animal forms that intelligence resides in today.

2. Old-Brain Equivalent

To create an intelligent machine, a few things are needed that exist in the older parts of our brain. Earlier, I said we don't need to replicate the old-brain areas. That is true in general, but there are some things the old brain does that are requirements for intelligent machines.

One requirement is basic movements. Recall that the neocortex does not directly control any muscles. When the neocortex wants to do something, it sends signals to older parts of the brain that more directly control movements. For example, balancing on two feet, walking, and running are behaviors implemented by older parts of the brain. You don't rely on your neocortex to balance, walk, and run. This makes sense, since animals needed to walk and run long before we evolved a neocortex. And why would we want the neocortex thinking about every step when it could be thinking about which path to take to escape a predator?

But does it have to be this way? Couldn't we build an intelligent machine where the neocortex equivalent directly controlled movements? I don't think so. The neocortex implements a near universal algorithm, but this flexibility comes with a price. The neocortex must be attached to something that already has sensors and already has behaviors. It does not create completely new behaviors; it learns how to string together existing ones in new and useful ways. The behavioral primitives can be as simple as the flexing of a finger or as complex as walking, but the neocortex requires that they exist. The behavioral primitives in the older parts of the brain are not all fixed—they can also be modified with learning. Therefore, the neocortex must continually adjust as well.

Behaviors that are intimately tied to the embodiment of a machine should be built in. For example, say we have a flying drone whose purpose is to deliver emergency supplies to people suffering from a natural disaster. We might make the drone intelligent, letting it assess on its own what areas are most in need and letting it coordinate with other drones when delivering its supplies. The "neocortex" of the drone cannot control all aspects of flight, nor would we want it to. The drone should have built-in behaviors for stable flight, landing, avoiding obstacles, etc. The intelligent part of the drone would not have to think about flight control in the same way that your neocortex does not have to think about balancing on two feet.

Safety is another type of behavior we should build into an intelligent machine. Isaac Asimov, the science-fiction writer, famously proposed three laws of robotics. These laws are like a safety protocol:

1. A robot may not injure a human being or, through inaction, allow a human being to come to harm.
2. A robot must obey orders given it by human beings except where such orders would conflict with the First Law.
3. A robot must protect its own existence as long as such protection does not conflict with the First or Second Law.

Asimov's three laws of robotics were proposed in the context of science-fiction novels and don't necessarily apply to all forms of machine intelligence. But in any product design, there are safeguards that are worth considering. They can be quite simple. For example, my car has a built-in safety system to avoid accidents. Normally, the car follows my orders, which I communicate via the accelerator and brake pedals. However, if the car detects an obstacle that I am going to hit, it will ignore my orders and apply the brakes. You could say the car is following Asimov's first and second laws, or you could say that the engineers who designed my car built in some safety features. Intelligent machines will also have built-in behaviors for safety. I include this idea here for complete-ness, even though these requirements are not unique to intelligent machines.

Finally, an intelligent machine must have goals and moti-vations. Human goals and motivations are complex. Some are driven by our genes, such as the desire for sex, food, and shelter. Emotions—such as fear, anger, and jealousy—can also have a large influence on how we behave. Some of our goals and motivations are more societal. For example, what is viewed as a successful life varies from culture to culture.

Intelligent machines also need goals and motivations. We wouldn't want to send a team of robotic construction workers to Mars, only to find them lying around in the sunlight all day charging their batteries. So how do we endow an intelligent ma-chine with goals, and is there a risk in this?

First, it is important to remember that the neocortex, on its own, does not create goals, motivations, or emotions. Recall the analogy I made between the neocortex and a map of the world. A map can tell us how to get from where we are to where we want to be, what will happen if we act one way or another, and what things are lo-cated at various places. But a map has no motivations on its own. A map will not desire to go someplace, nor will it spontaneously develop goals or ambitions. The same is true for the neocortex.

The neocortex is actively involved in how motivations and goals influence behavior, but the neocortex does not lead. To get a sense of how this works, imagine older brain areas conversing with the neocortex. Old brain says, "I am hungry. I want food." The neo-cortex responds, "I looked for food and found two places nearby that had food in the past. To reach one food location, we follow a river. To reach the other, we cross an open field where some tigers live." The neocortex says these things calmly and without value. However, the older brain area associates tigers with danger. Upon hearing the word "tiger," the old brain jumps into action. It releases chemicals into the blood that raise your heart rate and causes other physiological effects that we associate with fear. The old brain may also release chemicals, called neuromodulators, directly into broad areas of the neocortex—in essence, telling the neocortex, "Whatever you were just thinking, DON'T do that."

To endow a machine with goals and motivations requires that we design specific mechanisms for goals and motivations and then embed them into the embodiment of the machine. The goals could be fixed, like our genetically determined desire to eat, or they could be learned, like our societally determined goals for how to live a good life. Of course, any goals must be built on top of safety measures such as Asimov's first two laws. In summary, an intelligent machine will need some form of goals and motivations; however, goals and motivations are not a consequence of intelligence, and will not appear on their own.

3. Neocortex Equivalent

The third ingredient for an intelligent machine is a general-purpose learning system that performs the same functions as the neocortex. Once again, there can be a wide range of design options. I will discuss two: speed and capacity.

Speed

Neurons take at least five milliseconds to do anything useful. Transistors made of silicon can operate almost a million times faster. Thus, a neocortex made of silicon could potentially think and learn a million times faster than a human. It is hard to imagine what such a dramatic improvement in speed of thought would lead to. But before we let our imaginations run wild, I need to point out that just because part of an intelligent machine can operate a million times faster than a biological brain doesn't mean the entire intelligent machine can run a million times faster, or that knowledge can be acquired at that speed.

For example, recall our robotic construction workers, the ones we sent to Mars to build a habitat for humans. They might be able to think and analyze problems quickly, but the actual process of construction can only be sped up a little bit. Heavy materials can be moved only so fast before the forces involved cause them to bend and break. If a robot needs to drill a hole in a piece of metal, it will happen no faster than if a human were drilling the hole. Of course, the robot construction workers might work continuously, not get tired, and make fewer mistakes. So, the entire process of preparing Mars for humans might occur several times faster when using intelligent machines compared to humans, but not a million times faster.

Consider another example: What if we had intelligent machines that did the work of neuroscientists, only the machines could think a million times faster? Neuroscientists have taken decades to reach our current level of understanding of the brain. Would that progress have occurred a million times faster, in less than an hour, with AI neuroscientists? No. Some scientists, like me and my team, are theorists. We spend our days reading papers, debating possible theories, and writing software. Some of this work could, in principle, occur much faster if performed by an intelligent machine. But our software simulations would still take days to run. Plus, our theories are not developed in a vacuum; we are dependent upon experimental discoveries. The brain theory in this book was constrained and informed by the results from hundreds of experimental labs. Even if we were able to think a million times faster, we would still have to wait for the experimentalists to publish their results, and they cannot significantly speed up their experiments. For example, rats have to be trained and data collected. Rats can't be sped up by any amount. Once again, using intelligent machines instead of humans to study neuroscience would speed up the rate of scientific discovery, but not by a million times.

Neuroscience is not unique in this regard. Almost all fields of scientific inquiry rely on experimental data. For example, today there are numerous theories about the nature of space and time. To know if any of these theories are correct requires new experimental data. If we had intelligent machine cosmologists that thought a million times faster than human cosmologists, they might be able to quickly generate new theories, but we would still have to build space telescopes and underground particle detectors to collect the data needed to know if any of the theories are correct. We can't dramatically speed up the creation of telescopes and particle detectors, nor can we reduce the time it takes for them to collect data.

There are some fields of inquiry that could be sped up significantly. Mathematicians mostly think, write, and share ideas. In principle, intelligent machines could work on some math problems a million times faster than human mathematicians. Another example is our virtual intelligent machine that crawls around the internet. The speed at which the intelligent Web crawler can learn is restricted by how quickly it can "move" by following links and opening files. This could be very fast.

Today's computers are probably a good analogy for what we can expect to happen. Computers do tasks that humans used to do by hand, and they do them about a million times faster. Computers have changed our society and have led to a dramatic increase in our ability to make scientific and medical discoveries. But computers have not led to a millionfold increase in the rate at which we do these things. Intelligent machines will have a similar impact on our society and how fast we make discoveries.

Capacity

Vernon Mountcastle realized that our neocortex got large, and we got smarter, by making copies of the same circuit, the cortical column. Machine intelligence can follow the same plan. Once we fully understand what a column does and how to make one out of silicon, then it should be relatively easy to create intelligent machines of varying capacity by using more or fewer column elements.

There aren't any obvious limits to how big we can make artificial brains. A human neocortex contains about 150,000 columns. What would happen if we made an artificial neocortex with 150 million? What would be the benefit of a brain one thousand times bigger than a human brain? We don't know yet, but there are a few observations that are worth sharing.

The size of neocortical regions varies considerably between people. For example, region V1, the primary visual region, can be twice as big in some people as in others. V1 is the same thickness for everyone, but the area, and hence the number of columns, can vary. A person with a relatively small V1 and a person with a relatively large V1 both have normal vision and neither person is aware of the difference. There is a difference, however; a person with a large V1 has higher acuity, meaning they can see smaller things. This might be useful if you were a watchmaker, for example. If we generalize from this, then increasing the size of some regions of the neocortex can make a modest difference, but it doesn't give you some superpower.

Instead of making regions bigger, we could create more regions and connect them in more complex ways. To some extent, this is the difference between monkeys and humans. A monkey's visual ability is similar to a human's, but humans have a bigger neocortex overall, with more regions. Most people would agree that a human is more intelligent than a monkey, that our model of the world is deeper and more comprehensive. This suggests that intelligent machines could surpass humans in the depth of their understanding. This doesn't necessarily mean that humans couldn't understand what an intelligent machine learns. For example, even though I could not have discovered what Albert Einstein did, I can understand his discoveries.

There is one more way that we can think about capacity. Much of the volume of our brain is wiring, the axons and dendrites that connect neurons to each other. These are costly in terms of energy and space. To conserve energy, the brain is forced to limit the wiring and therefore limit what can be readily learned. When we are born, our neocortex has an overabundance of wiring. This is pared down significantly during the first few years of life. Presumably the brain is learning which connections are useful and which are not based on the early life experiences of the child. The removal of unused wiring has a downside, though; it makes it difficult to learn new types of knowledge later in life. For example, if a child is not exposed to multiple languages early in life, then the ability to become fluent in multiple languages is diminished. Similarly, a child whose eyes do not function early in life will permanently lose the ability to see, even if the eyes are later repaired. This is probably because some of the connections that are needed for being multilingual and for seeing were lost because they weren't being used.

Intelligent machines do not have the same constraints related to wiring. For example, in the software models of the neocortex that my team creates, we can instantly establish connections between any two sets of neurons. Unlike the physical wiring in the brain, software allows all possible connections to be formed. This flexibility in connectivity could be one of the greatest advantages of machine intelligence over biological intelligence. It could allow intelligent machines to keep all their options open, as it removes one of the greatest barriers human adults face when trying to learn new things.

Learning Versus Cloning

Another way that machine intelligence will differ from human intelligence is the ability to clone intelligent machines. Every human has to learn a model of the world from scratch. We start life knowing almost nothing and spend several decades learning. We go to school to learn, we read books to learn, and of course we learn via our personal experiences. Intelligent machines will also have to learn a model of the world. However, unlike humans, at any time we can make a copy of an intelligent machine, cloning it. Imagine we have a standardized hardware design for our intelligent Mars construction robots. We might have the equivalent of a school to teach a robot about construction methods, materials, and how to use tools. This training might take years to complete. But once we are satisfied with the robot's abilities, we can make copies by transferring its learned connections into a dozen other identical robots.

The next day we could reprogram the robots again, with an improved design or perhaps with entirely new skills.

The Future Applications of Machine Intelligence Are Unknown

When we create a new technology, we imagine that it will be used to replace or improve something we are familiar with. Over time, new uses arise that no one anticipated, and it is these unanticipated uses that typically become most important and transform society. For example, the internet was invented to share files between scientific and military computers, something that had previously been done manually but could now be done faster and more efficiently. The internet is still used to share files, but, more importantly, it radically transformed entertainment, commerce, manufacturing, and personal communication. It has even changed how we write and read. Few people imagined these societal shifts when the internet protocols were first created.

Machine intelligence will undergo a similar transition. Today, most AI scientists focus on getting machines to do things that humans can do—from recognizing spoken words, to labeling pictures, to driving cars. The notion that the goal of AI is to mimic humans is epitomized by the famous "Turing test." Originally proposed by Alan Turing as the "imitation game," the Turing test states that if a person can't tell if they are conversing with a computer or a human, then the computer should be considered intelligent. Unfortunately, this focus on human-like ability as a metric for intelligence has done more harm than good. Our excitement about tasks such as getting a computer to play Go has distracted us from imagining the ultimate impact of intelligent machines.

Of course, we will use intelligent machines to do things we humans do today. This will include dangerous and unhealthy jobs that are perhaps too risky for humans, such as deep-sea repair or cleaning up toxic spills. We will also use intelligent machines for tasks where there aren't enough humans, perhaps as caregivers for the elderly. Some people will want to use intelligent machines to replace good-paying jobs or to fight wars. We will have to work to find the right solutions to the dilemmas some of these applications will present.

But what can we say about the unanticipated applications of machine intelligence? Although no one can know the details of the future, we can try to identify large ideas and trends that might propel the adoption of AI in unanticipated directions. One that I find exciting is the acquisition of scientific knowledge. Humans want to learn. We are drawn to explore, to seek out knowledge, and to understand the unknown. We want to know the answers to the mysteries of the universe: How did it all begin? How will it end? Is life common in the universe? Are there other intelligent beings? The neocortex is the organ that allows humans to seek this knowledge. When intelligent machines can think faster and deeper than us, sense things we can't sense, and travel to places we can't travel to, who knows what we will learn. I find this possibility exciting.

Not everyone is as optimistic as I am about the benefits of machine intelligence. Some people see it as the greatest threat to humanity. I discuss the risks of machine intelligence in the next chapter.

CHAPTER 11

The Existential Risks of Machine Intelligence

At the beginning of the twenty-first century, the field of artificial intelligence was viewed as a failure. When we started Numenta, we conducted market research to see what words we might use to talk about our work. We learned that the terms "AI" and "artificial intelligence" were viewed negatively by almost everyone. No company would consider using them to describe their products. The general view was that attempts to build intelligent machines had stalled and might never succeed. Within ten years, people's impression of AI had flipped completely. It is now one of the hottest fields of research, and companies are applying the AI moniker to practically anything that involves machine learning.

Even more surprising was how quickly technology pundits changed from "AI might never happen" to "AI is likely to destroy all humans in the near future." Several nonprofit institutes and think tanks have been created to study the existential risks of AI, and numerous high-profile technologists, scientists, and philosophers have publicly warned that the creation of intelligent machines might rapidly lead to human extinction or subjugation.

Artificial intelligence is now viewed by many as an existential threat to humanity.

Every new technology can be abused to cause harm. Even today's limited AI is being used to track people, influence elections, and spread propaganda. These kinds of abuses will get worse when we have truly intelligent machines. For example, the idea that weapons can be made intelligent and autonomous is scary. Imagine intelligent drones that, instead of delivering medicines and food, are delivering weapons. Because intelligent weapons can act without human supervision, they could be deployed by the tens of thousands. It is essential that we confront these threats and institute policies to prevent bad outcomes.

Bad people will try to use intelligent machines to take away freedoms and threaten lives, but for the most part, a person using intelligent machines for bad purposes would not likely lead to the extermination of all humans. Concerns about the existential risks of AI, on the other hand, are qualitatively different. It is one thing for bad people to use intelligent machines to do bad things; it is something else if the intelligent machines are themselves bad actors and decide on their own to wipe out humanity. I am going to focus only on the latter possibility, the existential threats of AI. By doing so, I don't intend to diminish the significant risks of people misusing AI.

The perceived existential risks of machine intelligence are largely based on two concerns. The first is called the intelligence explosion. The story goes like this: We create machines that are more intelligent than humans. These machines are better than humans at pretty much everything, including creating intelligent machines. We let the intelligent machines create new intelligent machines, which then create machines that are even more intelligent. The time between each improved generation of intelligent machine gets shorter and shorter, and before long the machines outdistance our intelligence by so much that we can't understand what they are doing. At this point, the machines might decide to get rid of us because they no longer need us (human extinction), or they might decide to tolerate us because we are useful to them (human subjugation).

The second existential risk is called goal misalignment, which refers to scenarios where intelligent machines pursue goals that are counter to our well-being and we can't stop them. Technologists and philosophers have posited several ways that this might happen. For example, intelligent machines might spontaneously develop their own goals that are harmful to us. Or they might pursue a goal we assigned to them, but do it so ruthlessly that they consume all of Earth's resources and, in the process, make the planet uninhabitable for us.

The underlying assumption of all these risk scenarios is that we lose control of our creations. Intelligent machines prevent us from turning them off or from stopping them in other ways from pursuing their goals. Sometimes it is assumed that the intelligent machines replicate, creating millions of copies of themselves, and in other scenarios a single intelligent machine becomes omnipotent. Either way, it is us versus them, and the machines are smarter.

When I read about these concerns, I feel that the arguments are being made without any understanding of what intelligence is. They feel wildly speculative, based on incorrect notions not just of what is technically possible, but what it means to be intelligent. Let's see how these concerns hold up when we consider them in light of what we have learned about the brain and biological intelligence.

The Intelligence Explosion Threat

Intelligence requires having a model of the world. We use our world model to recognize where we are and to plan our movements. We use our model to recognize objects, to manipulate them, and to anticipate the consequences of our actions. When we want to accomplish something, whether it is as simple as making a pot of coffee or as complex as overturning a law, we use the model in our brain to decide which actions we should take to reach the desired outcome.

With few exceptions, learning new ideas and skills requires physically interacting with the world. For example, the recent discoveries of planets in other solar systems required first building a new type of telescope and then collecting data over several years. No brain, no matter how big or fast, could know the prevalence and composition of extrasolar planets by just thinking. It is not possible to skip the observation phase of discovery. Learning how to fly a helicopter requires understanding how subtle changes in your behavior cause subtle changes in flight. The only way to learn these sensory-motor relationships is by practicing. Perhaps a machine could practice on a simulator, which in theory could be faster than learning by flying a real helicopter, but it would still take time. To run a factory making computer chips requires years of practice. You can read a book about chip manufacturing, but an expert has learned the subtle ways things can go wrong in the manufacturing process and how to address them. There is no substitute for this experience.

Intelligence is not something that can be programmed in software or specified as a list of rules and facts. We can endow a machine with the ability to learn a model of the world, but the knowledge that makes up that model has to be learned, and learning takes time. As I described in the previous chapter, although we can make intelligent machines that run a million times faster than a biological brain, they cannot acquire new knowledge a million times faster.

Acquiring new knowledge and skills takes time regardless of how fast or big a brain might be. In some domains, such as mathematics, an intelligent machine could learn much faster than a human. In most fields, however, the speed of learning is limited by the need to physically interact with the world. Therefore, there can't be an explosion of intelligence where machines suddenly know much more than we do.

The intelligence explosion adherents sometimes talk about "superhuman intelligence," which is when machines surpass human performance in every way and on every task. Think about what that implies. A superhuman intelligent machine could expertly fly every type of airplane, operate every type of machine, and write software in every programming language. It would speak every language, know the history of every culture in the world, and understand the architecture in every city. The list of things that humans can do collectively is so large that no machine can surpass human performance in every field.

Superhuman intelligence is also impossible because what we know about the world is constantly changing and expanding. For example, imagine that some scientists discover a new means of quantum communication, which allows instant transmission across vast distances. At first, only the humans who made this discovery would know about it. If the discovery was based on an experimental result, no one—and no machine, no matter how smart—could have just thought of it. Unless you assume that machines have replaced all the scientists in the world (and all the human experts in every field), then some humans will always be more expert at some things than machines. This is the world we live in today. No human knows everything. This isn't because nobody is smart enough; it is because no one person can be everywhere and do everything. The same is true for intelligent machines.

Notice that most of the successes of current AI technology are on problems that are static—that don't change over time and don't require continuous learning. For example, the rules of Go are fixed. The mathematical operations that my calculator performs do not change. Even systems that label images are trained and tested using a fixed set of labels. For static tasks such as these, a dedicated solution can not only outperform humans, but do so indefinitely. However, most of the world is not fixed, and the tasks we need to perform are constantly changing. In such a world, no human or machine can have a permanent advantage on any task, let alone all tasks.

People who worry about an intelligence explosion describe intelligence as if it can be created by an as-yet-to-be-discovered recipe or secret ingredient. Once this secret ingredient is known, it can be applied in greater and greater quantities, leading to superintelligent machines. I agree with the first premise. The secret ingredient, if you will, is that intelligence is created through thousands of small models of the world, where each model uses reference frames to store knowledge and create behaviors. However, adding this ingredient to machines does not impart any immediate capabilities. It only provides a substrate for learning, endowing machines with the ability to learn a model of the world and thus acquire knowledge and skills. On a kitchen stovetop you can turn a knob to up the heat. There isn't an equivalent knob to "up the knowledge" of a machine.

The Goal-Misalignment Threat

This threat supposedly arises when an intelligent machine pursues a goal that is harmful to humans *and* we can't stop it. It is sometimes referred to as the "Sorcerer's Apprentice" problem. In the story by Goethe, a sorcerer's apprentice enchants a broom to fetch water, but then realizes that he doesn't know how to stop the broom from fetching water. He tries cutting the broom with an ax, which only leads to more brooms and more water. The concern is that an intelligent machine might similarly do what we ask it to do, but when we ask the machine to stop, it sees that as an obstacle to completing the first request. The machine goes to any length to pursue the first goal. One commonly discussed illustration of the goal-misalignment problem is asking a machine to maximize the production of paper clips. Once the machine starts pursuing this task, nothing can stop it. It turns all the Earth's resources into paper clips.

The goal-misalignment threat depends on two improbabilities: first, although the intelligent machine accepts our first request, it ignores subsequent requests, and, second, the intelligent machine is capable of commandeering sufficient resources to prevent all human efforts to stop it.

As I have pointed out multiple times, intelligence is the ability to learn a model of the world. Like a map, the model can tell you how to achieve something, but on its own it has no goals or drives. We, the designers of intelligent machines, have to go out of our way to design in motivations. Why would we design a machine that accepts our first request but ignores all others after that? This is as likely as designing a self-driving car that, once you tell it where you want to go, ignores any further requests to stop or go someplace else. Further, it assumes that we designed the car so that it locks all the doors and disconnects the steering wheel, brake pedal, power button, etc. Note that a self-driving car isn't going to develop goals on its own. Of course, someone could design a car that pursues its own goals and ignores the requests of humans. Such a car might cause harm. But even if someone did design such a machine, it would not be an existential threat without satisfying the second requirement.

The second requirement of the goal-misalignment risk is that an intelligent machine can commandeer the Earth's resources to pursue its goals, or in other ways prevent us from stopping it. It is difficult to imagine how this could happen. To do so would require the machine to be in control of the vast majority of the world's communications, production, and transportation. Clearly, a rogue intelligent car can't do this. A possible way for an intelligent machine to prevent us from stopping it is blackmail. For example, if we put an intelligent machine in charge of nuclear weapons, then the machine could say, "If you try to stop me, I will blow us all up." Or if a machine controlled most of the internet, it could threaten all sorts of mayhem by disrupting communications and commerce.

We have similar concerns with humans. This is why no single person or entity can control the entire internet and why we require multiple people to launch a nuclear missile. Intelligent machines will not develop misaligned goals unless we go to great lengths to endow them with that ability. Even if they did, no machine can commandeer the world's resources unless we let it. We don't let a single human, or even a small number of humans, control the world's resources. We need to be similarly careful with machines.

The Counterargument

I am confident that intelligent machines do not pose an existential threat to humanity. A common counterargument from those who disagree goes like this: Indigenous people throughout history felt similarly secure. But when foreigners showed up with superior weapons and technology, the indigenous populations were overcome and destroyed. We are, they argue, similarly vulnerable and can't trust our sense of security. We cannot imagine how much smarter, faster, and more capable machines might be compared to us, and therefore we are vulnerable.

There is some truth to this argument. Some intelligent machines will be smarter, faster, and more capable than humans. The issue of concern circles back to motivation. Will intelligent machines want to take over Earth, or subjugate us, or do anything that might harm us? The destruction of indigenous cultures arose from the motivations of the invaders, which included greed, fame, and a desire for dominance. These are old-brain drives. Superior technology helped the invaders, but it wasn't the root cause of the carnage.

Once again, intelligent machines will not have human-like emotions and drives unless we purposely put them there. Desires, goals, and aggression do not magically appear when something is intelligent. To support my point, consider that the largest loss of indigenous life was not directly inflicted by human invaders, but by introduced diseases—bacteria and viruses for which indigenous people had poor or no defenses. The true killers were simple organisms with a drive to multiply and no advanced technology. Intelligence had an alibi; it wasn't present for the bulk of the genocide.

I believe that self-replication is a far greater threat to humanity than machine intelligence. If a bad person wanted to create something to kill all humans, the surer way to do this would be to design new viruses and bacteria that are highly infectious and that our immune systems can't defend against. It is theoretically possible that a rogue team of scientists and engineers could design intelligent machines that wanted to self-replicate. The machines would also need to be able to make copies of themselves without the possible interference of humans. These events seem highly unlikely, and even if they happened, none of it would occur quickly. The point is that anything capable of self-replication, especially viruses and bacteria, is a potential existential threat. Intelligence, on its own, is not.

We can't know the future, and therefore we can't anticipate all the risks associated with machine intelligence, just as we can't anticipate all the risks for any other new technology. But as we go forward and debate the risks versus the rewards of machine intelligence, I recommend acknowledging the distinction between three things: replication, motivations, and intelligence.

- **Replication**: Anything that is capable of self-replication is dangerous. Humanity could be wiped out by a biological virus. A computer virus could bring down the internet. Intelligent machines will not have the ability or the desire to self-replicate unless humans go to great lengths to make it so.
- **Motivations**: Biological motivations and drives are a consequence of evolution. Evolution discovered that animals with certain drives replicated better than other animals. A machine that is not replicating or evolving will not suddenly develop a desire to, say, dominate or enslave others.
- **Intelligence**: Of the three, intelligence is the most benign. An intelligent machine will not on its own start to self-replicate, nor will it spontaneously develop drives and motivations. We will have to go out of our way to design in the motivations we want intelligent machines to have. But unless intelligent

machines are self-replicating and evolving, they will not, on their own, represent an existential risk to humanity.

I don't want to leave you with the impression that machine intelligence is not dangerous. Like any powerful technology, it could cause great harm if used by humans with ill intent. Again, just imagine millions of intelligent autonomous weapons or using intelligent machines for propaganda and political control. What should we do about this? Should we impose a ban on AI research and development? That would be difficult, but it might also be contrary to our best interests. Machine intelligence will greatly benefit society and, as I will argue in the next section of the book, it might be necessary for our long-term survival. For now, it appears that our best option is to work hard to form enforceable international agreements on what is acceptable and what is not, similar to how we treat chemical weapons.

Machine intelligence is often compared to a genie in a bottle. Once released, it cannot be put back, and we will quickly lose our ability to control it. What I hoped to show in this chapter is that these fears are ungrounded. We will not lose control, and nothing is going to happen rapidly, as the proponents of an intelligence explosion fear. If we start now, we have plenty of time to sort through the risks and rewards and decide how we want to go forward.

In the next and final section of the book, we look at the existential risks and opportunities of human intelligence.

PART 3

Human Intelligence

We are at an inflection point in the history of Earth, a period of rapid and dramatic change to both the planet and the life-forms that inhabit it. The climate is changing so rapidly that it is likely to make some cities uninhabitable and large agricultural areas barren in the next one hundred years. Species are going extinct at such a rapid rate that some scientists are calling it the sixth great extinction event in Earth's history. Human intelligence is the cause of these rapid changes.

Life appeared on Earth about 3.5 billion years ago. From the beginning, the course that life took was ruled by genes and evolution. There is no plan or desired direction to evolution. Species evolved and went extinct based on their ability to leave offspring with copies of their genes. Life was driven by competitive survival and procreation. Nothing else mattered.

Our intelligence has allowed our species, Homo sapiens, to flourish and succeed. In just a couple hundred years—a near instant in geological time—we have doubled our life expectancy, cured many diseases, and eliminated hunger for the vast majority of humans. We live healthier, are more comfortable, and toil less than our predecessors.

Humans have been intelligent for hundreds of thousands of years, so why the sudden change in our fortunes? What is new is the recent and rapid rise of our technology and scientific discoveries, which have allowed us to produce food in abundance, eliminate diseases, and transport goods to wherever they are most needed.

But with our success, we have created problems. Our population has gone from one billion, two hundred years ago, to close to eight billion today. There are so many of us that we are polluting every part of the planet. It is now evident that our ecological impact is so severe that, at a minimum, it will displace hundreds of millions of people; at the worst, it will make Earth uninhabitable. Climate is not our only concern. Some of our technologies, such as nuclear weapons and gene editing, afford the potential for a small number of people to kill billions of others.

Our intelligence has been the source of our success, but it has also become an existential threat. How we act in the coming years will determine whether our sudden rise leads to a sudden collapse—or, alternately, if we exit this period of rapid change on a sustainable trajectory. These are the themes I discuss in the remaining chapters of the book.

I start by looking at the inherent risks associated with our intelligence and how our brains are structured. From this base, I discuss various options we might pursue that will increase the chances of our long-term survival. I discuss existing initiatives and proposals, looking at them through the lens of brain theory. And I discuss new ideas that I think should be considered, but, as far as I know, have not entered the mainstream discourse.

My goal is not to prescribe what we should do but to encourage conversations about issues that I believe are not being sufficiently discussed. Our new understanding of the brain affords the opportunity to take a fresh look at the risks and opportunities we face. Some of what I talk about might be a bit controversial, but that is not my intent. I am trying to provide an honest and unbiased assessment of the situation we find ourselves in, and to explore what we might do about it.

CHAPTER 12

False Beliefs

As teenagers, my friends and I were fascinated by the brain-in-a-vat hypothesis. Was it possible that our brains are sitting in a vat of nutrients that keep it alive, while the inputs and outputs are connected to a computer? The brain-in-a-vat hypothesis suggests the possibility that the world we think we live in might not be the real world, but instead a fake world simulated by a computer. Although I don't believe our brains are connected to a computer, what *is* happening is almost as strange. The world we think we live in is not the real world; it is a simulation of the real world. This leads to a problem. What we believe is often not true.

Your brain is in a box, the skull. There are no sensors in the brain itself, so the neurons that make up your brain are sitting in the dark, isolated from the world outside. The only way your brain knows anything about reality is through the sensory nerve fibers that enter the skull. The nerve fibers coming from the eyes, ears, and skin look the same, and the spikes that travel along them are identical. There is no light or sound entering the skull, only electrical spikes.

The brain also sends nerve fibers to the muscles, which move the body and its sensors and thereby change what part of the world the brain is sensing. By repeatedly sensing and moving, sensing and moving, your brain learns a model of the world outside the skull.

Notice again that there is no light, touch, or sound entering the brain. None of the perceptions that make up our mental experiences—from the fuzziness of a pet, to the sigh of a friend, to the colors of fall leaves—come through the sensory nerves. The nerves only send spikes. And since we do not perceive spikes, everything we do perceive must be fabricated in the brain. Even the most basic feelings of light, sound, and touch are creations of the brain; they only exist in its model of the world.

You might object to this characterization. After all, don't the input spikes *represent* light and sound? Sort of. There are properties of the universe, such as electromagnetic radiation and compression waves of gaseous molecules, that we can sense. Our sensory organs convert these properties into nerve spikes, which are then converted into our perception of light and sound. But the sensory organs do not sense everything. For example, light in the real world exists over a broad range of frequencies, but our eyes are only sensitive to a tiny sliver of this range. Similarly, our ears only detect sounds in a narrow range of audio frequencies. Therefore, our perception of light and sound can only represent part of what is going on in the universe. If we could sense all frequencies of electromagnetic radiation, then we would see radio broadcasts and radar and would have X-ray vision. With different sensors, the same universe would lead to different perceptual experiences.

The two important points are that the brain only knows about a subset of the real world, and that what we perceive is our model of the world, not the world itself. In this chapter, I explore how these ideas lead to false beliefs, and what, if anything, we can do about it.

We Live in a Simulation

At any moment in time, some of the neurons in the brain are active and some are not. The active neurons represent what we are currently thinking and perceiving. Importantly, these thoughts and perceptions are relative to the brain's model of the world, not the physical world outside the skull. Therefore, the world we perceive is a simulation of the real world.

I know it doesn't feel as if we are living in a simulation. It feels as if we are looking directly at the world, touching it, smelling it, and feeling it. For example, it is common to think the eyes are like a camera. The brain receives a picture from the eyes, and that picture is what we see. Although it is natural to think this way, it isn't true. Recall that earlier in the book I explained how our visual perception is stable and uniform, even though the inputs from the eyes are distorted and changing. The truth is, we perceive our model of the world, not the world itself or the rapidly changing spikes entering the skull. As we go about our day, the sensory inputs to the brain invoke the appropriate parts of our world model, but what we perceive and what we believe is happening is the model. Our reality is similar to the brain-in-a-vat hypothesis; we live in a simulated world, but it is not in a computer—it is in our head.

This is such a counterintuitive idea that it is worth going through several examples. Let's start with the perception of location. A nerve fiber representing pressure on a fingertip does not convey any information about where the finger is. The fingertip nerve fiber responds the same way whether your finger is touching something in front of you or off to your side. Yet you perceive the sense of touch as being at some location relative to your body. This seems so natural that you probably never asked how it occurs. As I discussed earlier, the answer is that there are cortical columns that represent each part of your body. And in those columns are neurons that represent the location of that body part. You perceive your finger to be someplace because the cells that represent the location of your finger say so.

The model can be wrong. For example, people who lose a limb often perceive that the missing limb is still there. The brain's model includes the missing limb and where it is located. So even though the limb no longer exists, the sufferer perceives it and feels that it is still attached. The phantom limb can "move" into different positions. Amputees may say that their missing arm is at their side, or that their missing leg is bent or straight. They can feel sensations, such as an itch or pain, located at particular locations on the limb. These sensations are "out there" where the limb is perceived to be, but, physically, nothing is there. The brain's model includes the limb, so, right or wrong, that is what is perceived.

Some people have the opposite problem. They have a normal limb but feel as if it doesn't belong to them. Because it feels alien, they may want to have the limb removed. Why some people feel that a limb does not belong to them is unknown, but the false perception is certainly rooted in their model of the world not having a normal representation for the limb. If your brain's model of your body does not include a left leg, then that leg will be perceived as not part of your body. It would be like someone gluing a coffee cup to your elbow. You would want to remove it as soon as you could.

Even a completely normal person's perception of their body can be fooled. The rubber hand illusion is a party game where the subject can see a rubber hand but not their real hand. When someone else identically strokes the rubber hand and the real, obscured hand, the subject will start to perceive that the rubber hand is actually part of their body.

These examples tell us that our model of the world can be incorrect. We can perceive things that don't exist (such as the phantom limb), and we can incorrectly perceive things that do exist (such as the alien limb and the rubber hand). These are examples where the brain's model is clearly wrong, and in a detrimental way. For example, phantom limb pain can be debilitating. Nevertheless, it is not uncommon for the brain's model to disagree with the brain's inputs. In most cases, this is useful.

The following image, created by Edward Adelson, is a powerful example of the difference between the brain's model of the world (what you perceive) and what is sensed. In the figure on the left, the square labeled A appears darker than the square labeled B. However, the A and B squares are identical. You may be saying to yourself, "That isn't possible. A is definitely darker than B." But you would be wrong. The best way to verify that A and B are the same is by blocking off all other parts of the image so that only these two squares remain visible, and then you will see that A and B are the identical shade. To help you, I include two clips from the main image. The effect is less pronounced in the slice and completely missing when only seeing the A and B squares.

To call this an illusion is to suggest that the brain is being tricked, but the opposite is true. Your brain is correctly perceiving a checkerboard and not being fooled by the shadow. A checkerboard pattern is a checkerboard pattern, regardless of whether it has a shadow on it. The brain's model says that checkerboard patterns have alternating dark and light squares, so that is what you perceive, even though in this instance the light coming from a "dark" square and a "light" square is identical.

The model of the world that resides in our brain is usually accurate. It usually captures the structure of reality independent of our current point of view and other conflicting data, such as the shadow on the checkerboard. However, the brain's model of the world can also be flat-out wrong.

False Beliefs

A false belief is when the brain's model believes that something exists that does not exist in the physical world. Think about phantom limbs again. A phantom limb occurs because there are columns in the neocortex that model the limb. These columns have neurons that represent the location of the limb relative to the body. Immediately after the limb is removed, these columns are still there, and they still have a model of the limb. Therefore, the sufferer believes the limb is still in some pose, even though it does not exist in the physical world. The phantom limb is an example of a false belief. (The perception of the phantom limb typically disappears over a few months as the brain adjusts its model of the body, but for some people it can last years.)

Now consider another false model. Some people believe that the world is flat. For tens of thousands of years, every human's experiences were consistent with the world being flat. The curvature of Earth is so slight that, through the course of a life, it was not possible to detect it. There are a few subtle inconsistencies, such as how the hull of a ship disappears over the horizon before the masts, but this is hard to see even with excellent eyesight. A model that says the Earth is flat is not only consistent with our sensations but a good model for acting in the world. For example, today I need to walk from my office to the library to return a book. Planning my trip to the library using a flat-Earth model works well; I don't have to consider the curvature of Earth to move about town. In terms of everyday survival, a flat-Earth model is a perfectly good one, or at least it was until recently. Today, if you are an astronaut, or a ship pilot, or even a frequent international traveler, believing that the Earth is flat can have serious and deadly consequences. If you are not a long-distance traveler, then a flat-Earth model still works well for day-to-day living.

Why do some people still believe the Earth is flat? How do they maintain their flat-Earth model in the presence of contrary sensory input, such as pictures of Earth from space or accounts of explorers who have traversed across the South Pole?

Recall that the neocortex is constantly making predictions. Predictions are how the brain tests whether its model of the world is correct; an incorrect prediction indicates something is wrong with the model and needs to be fixed. A prediction error causes a burst of activity in the neocortex, which directs our attention to the input that caused the error. By attending to the mis-predicted input, the neocortex relearns that part of the model. This ultimately leads to a modification of the brain's model to more accurately reflect the world. Model repair is built into the neocortex, and normally it works reliably.

To hold on to a false model, such as a flat Earth, requires dismissing evidence that conflicts with your model. Flat-Earth believers say they distrust all evidence that they cannot directly sense. A picture can be fake. An explorer's account can be fabricated. Sending people to the moon in the 1960s could have been a Hollywood production. If you limit what you believe to only things you can directly experience, and you are not an astronaut, then a flat-Earth model is what you will end up with. To maintain a false model, it also helps to surround yourself with other people who have the same false beliefs, thus making it more likely that the inputs you receive are consistent with your model. Historically, this entailed physically isolating yourself in a community of people with similar beliefs, but today you can achieve a similar result by selectively watching videos on the internet.

Consider climate change. There is overwhelming evidence that human activity is leading to large-scale changes in the Earth's climate. These changes, if not checked, might lead to the death and/or displacement of billions of people. There are legitimate debates about what we should do about climate change, but there are many people who simply deny it is happening. Their model of the world says that the climate is not changing—or, even if it is changing, there is nothing to be concerned about.

How do climate-change deniers maintain their false belief in the face of substantial physical evidence? They are like flat-Earth believers: they don't trust most other people, and they rely only on what they personally observe or what other similarly minded people tell them. If they can't see the climate changing, then they don't believe it is happening. Evidence suggests that climate-change deniers are likely to become climate-change believers if they personally experience an extreme weather event or flooding due to rising seas.

If you rely only on your personal experiences, then it is possible to live a fairly normal life and believe that the Earth is flat, that the moon landings were faked, that human activity is not changing the global climate, that species don't evolve, that vaccines cause diseases, and that mass shootings are faked.

Viral World Models

Some models of the world are viral, by which I mean that the model causes the host brain to act in ways that spread the model to other brains. A model of a phantom limb is not viral; it is an incorrect model, but it is isolated to one brain. A flat-Earth model is also not viral, because to maintain it requires trusting only your personal experiences. Believing the Earth is flat does not cause you to act in a way that spreads your belief to other people.

Viral models of the world prescribe behaviors that spread the model from brain to brain in increasing numbers. For example, my world model includes the belief that every child should get a good education. If part of that education is to teach that every child deserves a good education, then this will inevitably lead to more and more people believing that every child deserves a good education. My model of the world, at least the part about universal childhood education, is viral. It will spread to more and more people over time. But is it correct? That is hard to say. My model of how humans ought to behave is not something physical, like the existence of a limb or the curvature of the Earth. Other people have a model that says only some children deserve a good education. Their model includes educating their children to believe that only they, and people like them, deserve a good education. This model of selective education is also viral, and arguably a better one for propagating genes. For example, the people who get a good education will get better access to financial resources and health care, and therefore be more likely to pass on their genes than those with little or no education. From a Darwinian point of view, selective education is a good strategy, as long as those not receiving an education don't rebel.

False and Viral World Models

Now we turn to the most troublesome type of world model: ones that are both viral and demonstrably false. For example, say we had a history book that contained many factual errors. The book begins with a set of instructions to the reader. The first instruction says, "Everything in this book is true. Ignore any evidence that contradicts this book." The second instruction says, "If you encounter others who also believe this book is true, then you should assist them in anything they need, and they will do the same for you." The third instruction says, "Tell everyone you can about the book. If they refuse to believe the book is true, then you should banish or kill them."

At first you might think, "Who is going to believe this?" However, if just a few people's brains believe the book is true, then brain models that include the book's veracity can spread virally to a great many other brains over time. The book not only describes a set of false beliefs about history but also prescribes specific actions. The actions cause people to spread belief in the book, help others who also believe in the book, and eliminate sources of contrary evidence.

The history book is an instance of a meme. First introduced by biologist Richard Dawkins, a meme is something that replicates and evolves, much like a gene, but through culture. (Recently, the term "meme" has been appropriated to represent images on the internet. I am using the word in its original definition.) The history book is actually a set of mutually supporting memes, in the same way that an individual organism is created by a set of mutually supporting genes. For example, each individual instruction in the book could be considered a meme.

The memes in the history book have a symbiotic relationship with the genes of a person who believes in the book. For example, the book dictates that people who believe in the book should get preferential support from other believers. This makes it likely that believers will have more surviving children (more copies of genes), which in turn leads to more people who believe the book is true (more copies of the memes).

Memes and genes evolve, and they can do so in a mutually reinforcing way. For example, let's say a variation of the history book is published. The difference between the original version and the new one is the addition of a few more instructions at the beginning of the book, such as "Women should have as many children as possible" and "Don't allow children to attend schools where they might be exposed to criticism of the book." Now there are two history books in circulation. The newer book, with its additional instructions, is slightly better at replication than the old book. So, over time, it will come to dominate. The biological genes of believers might similarly evolve to select people who are more willing to have many children, are better able to ignore evidence that contradicts the book, or are more willing to harm nonbelievers.

False models of the world can spread and thrive as long as the false beliefs help the believers spread their genes. The history book and the people who believe it are in a symbiotic relationship. They help each other replicate, and they evolve over time in a mutually reinforcing way. The history book may be factually incorrect, but life is not about having a correct model of the world. Life is about replication.

Language and the Spread of False Beliefs

Before language, an individual's model of the world was limited to places they had personally traveled and things they had personally encountered. No one could know what was over a ridge or across an ocean without going there. Learning about the world through personal experience is generally reliable.

With the advent of language, humans extended our model of the world to include things we have not personally observed. For example, even though I have never been to Havana, I can talk to people who claim to have been there and read what others have written about it. I believe Havana is a real place, because people I trust tell me that they have been there and their reports are consistent. Today, much of what we believe about the world is not directly observable, and therefore we rely on language to learn about these phenomena. This includes discoveries such as atoms, molecules, and galaxies. It includes slow processes such as the evolution of species and plate tectonics. It includes places we haven't personally traveled to but believe exist, such as the planet Neptune and, in my case, Havana. The triumph of the human intellect, the enlightenment of our species, is the expansion of our world model beyond what we can directly observe. This expansion of knowledge was made possible by tools—such as ships, microscopes, and telescopes—and by various forms of communication, such as written language and pictures.

But learning about the world indirectly through language is not 100 percent reliable. For example, it is possible that Havana is not a real place. It is possible that the people who told me about Havana are lying and coordinating their misinformation to fool me. The false history book shows how false beliefs can spread through language even if no one is intentionally spreading misinformation.

There is only one way, that we know of, to discern falsehoods from truths, one way to see if our model of the world has errors. That method is to actively seek evidence that contradicts our beliefs. Finding evidence that supports our beliefs is helpful, but not definitive. Finding contrary evidence, however, is proof that the model in our head is not right and needs to be modified. Actively looking for evidence to disprove our beliefs is the scientific method. It is the only approach we know of that can get us closer to the truth.

Today, in the beginning of the twenty-first century, false beliefs are rampant in the minds of billions of people. This is understandable for mysteries that have not yet been solved. For example, it is understandable that people believed in a flat Earth five hundred years ago, because the spherical nature of the planet was not widely understood, and there was little to no evidence that the Earth was not flat. Similarly, it is understandable that today there are different beliefs about the nature of time (all but one must be wrong), as we have not yet discovered what time is. But what is disturbing to me is that billions of people still hold beliefs that have been proven false. For example, three hundred years after the beginning of the Enlightenment, the majority of humans still believe in mythical origins of Earth. These origin myths have been proven false by mountains of contrary evidence, yet people still believe them.

We have viral false beliefs to blame for this. Like the fake history book, memes rely on brains to replicate, and therefore, they have evolved ways of controlling the behavior of brains to further their interests. Because the neocortex is constantly making predictions to test its model of the world, the model is inherently self-correcting. On its own, a brain will inexorably move toward more and more accurate models of the world. But this process is thwarted, on a global scale, by viral false beliefs.

Toward the end of the book, I will present a more optimistic view of humanity. But before we turn toward this brighter vision, I want to talk about the very real existential threat that we humans present to ourselves.

CHAPTER 13

The Existential Risks of Human Intelligence

Intelligence itself is benign. As I argued two chapters ago, unless we purposefully build in selfish drives, motivations, and emotions, intelligent machines will not pose a risk to our survival. Human intelligence, however, is not as benign. The possibility that human behavior might lead to our demise has been recognized for a long time. For example, since 1947 the *Bulletin of the Atomic Scientists* has maintained the Doomsday Clock to highlight how close we are to making Earth uninhabitable. First inspired by the possibility that a nuclear war and resulting conflagration could destroy Earth, the Doomsday Clock was expanded in 2007 to include climate change as a second potential cause of self-inflicted extinction. Whether nuclear weapons and human-induced climate change are existential threats is debated, but there is no question that both have the potential to cause great human suffering. With climate change, we are past any uncertainty; the debate has mostly shifted to how bad it will be, who will be affected, how quickly it will progress, and what we should do about it.

The existential threats of nuclear weapons and climate change did not exist one hundred years ago. Given the current rate of technological change, we will almost certainly create additional existential threats in the coming years. We need to fight these threats, but if we are to succeed in the long term then we need to look at these problems from a systemic point of view. In this chapter, I focus on the two fundamental systemic risks associated with the human brain.

The first is associated with the older parts of our brain. Although our neocortex endows us with superior intelligence, 30 percent of our brain evolved much longer ago and creates our more primitive desires and actions. Our neocortex has invented powerful technologies that are capable of changing the entire Earth, but the human behavior that controls these world-changing technologies is often dominated by the selfish and shortsighted old brain.

The second risk is more directly associated with the neocortex and intelligence. The neocortex can be fooled. It can form false beliefs about fundamental aspects of the world. Based on these false beliefs, we can act against our own long-term interests.

The Risks of the Old Brain

We are animals, descended from countless generations of other animals. Each and every one of our ancestors was successful in having at least one offspring, which in turn had at least one offspring, and so on. Our lineage goes back billions of years. Throughout this entire stretch of time, the ultimate measure of success—arguably the only one—was preferentially passing one's genes on to the next generation.

Brains were useful only if they increased the survival and fecundity of an animal possessing one. The first nervous systems were simple; they only controlled reflex reactions and bodily functions. Their design and function were completely specified by genes. Over time, the built-in functions expanded to include behaviors we consider desirable today, such as caring for offspring and social cooperation. But behaviors we look less kindly upon also appeared, such as fighting over territory, fighting for mating rights, forced copulation, and stealing resources.

All built-in behaviors, regardless of whether we think of them as desirable or not, came into being because they were successful adaptations. The older parts of our brain still harbor these primitive behaviors; we all live with this heritage. Of course, each of us lies somewhere along a spectrum of how much we express these old-brain behaviors and how much our more logical neocortex is able to control them. Some of this variation is believed to be genetic. How much is cultural is unknown.

So, even though we are intelligent, our old brain is still here. It is still operating under the rules laid down by hundreds of millions of years of survival. We still fight for territory, we still fight for mating rights, and we still cheat, rape, and trick our fellow humans. Not everyone does these things, and we teach our children the behaviors we want them to exhibit, but a quick look at any day's news will confirm that, as a species, across cultures and in every community, we haven't yet been able to free ourselves from these less desirable primitive behaviors. Again, when I refer to a behavior as being less desirable, I mean from an individual or societal point of view. From the perspective of genes, all these behaviors are useful.

On its own, the old brain doesn't represent an existential risk. The behaviors of the old brain are, after all, successful adaptations. In the past, if, in the pursuit of territory, one tribe killed all the members of another tribe, it didn't threaten all humans. There were winners as well as losers. The actions of one or a few people were limited to a part of the globe and a part of humanity. Today, the old brain represents an existential threat because our neocortex has created technologies that can alter and even destroy the entire planet. The shortsighted actions of the old brain, when paired with the globe-altering technologies of the neocortex, have become an existential threat to humanity. Let's look at how this plays out today by examining climate change and one of its underlying causes, population growth.

Population Growth and Climate Change

Human-caused climate change is a result of two factors. One is the number of people who live on Earth and the other is how much pollution each person creates. Both of these numbers are going up. Let's look at population growth.

In 1960, there were about three billion people on the planet. My earliest memories are from that decade. I don't recall anyone proposing that the problems the world faced in the 1960s could be solved if only we had twice as many people. Today, the world's human population is approaching eight billion and continuing to grow.

Simple logic says that the Earth would be less likely to experience some form of human-caused degradation and collapse if there were fewer people. For example, if there were two billion people instead of eight billion, then it is possible that the Earth's ecosystems could absorb our impact without rapid and radical change. Even if the Earth couldn't sustainably handle the presence of two billion humans, we would have more time to adjust our behaviors to live in a sustainable way.

Why, then, did the Earth's population go from three billion in 1960 up to eight billion today? Why didn't the population stay at three billion, or go down to two billion? Almost everyone would agree that the Earth would be better off with fewer people instead of more. Why isn't that happening? The answer may be obvious, but it is worth dissecting it a bit.

Life is based on a very simple idea: genes make as many copies of themselves as possible. This led to animals trying to have as many children as possible and to species trying to inhabit as many locales as possible. Brains evolved to serve this most basic aspect of life. Brains help genes make more copies of themselves.

However, what is good for genes is not always good for individuals. For example, from a gene's point of view, it is fine if a human family has more children than it can feed. Sure, in some years children may die of starvation, but in other years they won't.

From a gene's perspective, it is better to occasionally have too many children than too few. Some children will suffer horribly, and parents will struggle and grieve, but genes don't care. We, as individuals, exist to serve the needs of genes. Genes that lead us to have as many children as possible will be more successful, even if that sometimes leads to death and misery.

Similarly, from a gene's perspective, it is best if animals try to live in new locations, even if these attempts often fail. Say a human tribe splits and occupies four new habitats, but only one of the splinter groups survives while the other three struggle, starve, and ultimately die out. There will be much misery for the individual humans but success for the gene, as it now occupies twice as much territory as before.

Genes don't understand anything. They don't enjoy being genes, and they don't suffer when they fail to replicate. They are simply complex molecules that are capable of replication.

The neocortex, on the other hand, understands the larger picture. Unlike the old brain—with its hardwired goals and behaviors—the neocortex learns a model of the world and can predict the consequences of uncontrolled population growth. Thus, we can anticipate the misery and suffering that we will endure if we continue to let the number of people on Earth grow. So why aren't we collectively lowering the population? Because the old brain is still in charge.

Recall the example of a tempting piece of cake that I mentioned in Chapter 2. Our neocortex may know that eating cake is bad for us, that it can lead to obesity, disease, and early death. We may leave the house in the morning resolved to eat only healthy food. Yet, when we see and smell a piece of cake, we often eat it anyway. The old brain is in control, and the old brain evolved in a time when calories were hard to come by. The old brain does not know about future consequences. In the battle between the old brain and the neocortex, the old brain usually wins. We eat the cake.

Since we have difficulty controlling our eating, we do what we can. We use our intelligence to mitigate the damage. We create medical interventions, such as drugs and surgeries. We hold conferences on the epidemic of obesity. We create campaigns to educate people about the risks of bad food. But even though, logically, it would be best if we just ate better, the fundamental problem remains. We still eat cake.

Something analogous is happening with population growth. We know that at some point we will have to stop our population's growth. This is simple logic; populations cannot grow forever, and many ecologists believe ours is already unsustainable. But we find it difficult to manage our population because the old brain wants to have children. So instead we have used our intelligence to dramatically improve farming, inventing new crops and new methods to increase yield. We have also created technologies that allow us to ship food anywhere in the world. Using our intelligence, we have achieved the miraculous: we have reduced hunger and famine during a period when the human population has almost tripled. However, this can only go on for so long. Either population growth stops or sometime in the future there will be great human suffering on Earth. That is a certainty.

Of course, this situation is not as black-and-white as I have portrayed. Some people logically decide to have fewer or no children, some may not have the education to understand the long-term threats of their actions, and many are so poor that they rely on having children for survival. The issues related to population growth are complex, but if we step back and look at the big picture, we see that humans have understood the threat of population growth for at least fifty years, and in that time our population has nearly tripled. At the root of this growth are old-brain structures and the genes they serve. Fortunately, there are ways that the neocortex can win this battle.

How the Neocortex Can Thwart the Old Brain

The odd thing about overpopulation is that the idea of having a smaller human population isn't controversial, but talking about how we might achieve it from where we are today is socially and politically unacceptable. Perhaps we recall China's largely decried one-child policy. Perhaps we unconsciously associate reducing the population with genocide, eugenics, or pogroms. For whatever reason, purposefully aiming for a smaller population is rarely discussed. Indeed, when a country's population is declining, such as in Japan today, it is considered an economic crisis. It is rare to hear Japan's shrinking population described as a role model for the rest of the world.

We are fortunate that there is a simple and clever solution to population growth, a solution that does not force anyone to do anything they don't want to do, a solution that we know will reduce our population to a more sustainable size, and a solution that also increases the happiness and well-being of the people involved. But it is a solution that many people object to nonetheless. The simple and clever solution is to make sure every woman has the ability to control her own fertility and is empowered to exercise that option if she wants to.

I call this a clever solution because, in the battle between the old brain and the neocortex, the old brain almost always wins. The invention of birth control shows how the neocortex can use its intelligence to get the upper hand.

Genes propagate best when we have as many offspring as possible. The desire for sex is the mechanism that evolution came up with to serve the genes' interests. Even if we don't want more children, it is difficult to stop having sex. So, we used our intelligence to create birth-control methods that let the old brain have as much sex as it wants without creating more children. The old brain is not intelligent; it doesn't understand what it is doing or why. Our neocortex, with its model of the world, can see the downsides of having too many children, and it can see the benefits of delaying starting a family. Instead of fighting the old brain, the neocortex lets the old brain get what it wants but prevents the undesirable end result.

Why, then, is there sustained resistance to empowering women? Why do many people oppose equal pay, universal day care, and family planning? And why do women still find obstacles to attaining equal representation in positions of power? By almost every objective measure, empowering women will lead to a more sustainable world with less human suffering. From the outside looking in, it seems counterproductive to fight against this. We can blame the old brain and viral false beliefs for this dilemma. This brings us to the second fundamental risk of the human brain.

The Risk of False Beliefs

The neocortex, despite its amazing abilities, can be fooled. People are easily fooled into believing basic things about the world that are false. If you have false beliefs, then you might make fatally bad decisions. It can be especially bad if these decisions have global consequences.

I had my first exposure to the quandary of false beliefs in grade school. As I pointed out earlier, there are many sources of false beliefs, but this story relates to religions. One day during recess at the beginning of the school year, a group of about ten children gathered in a circle on the playground. I joined them. They were taking turns saying what religion they belonged to. As each kid stated what he or she believed, the other kids joined in to say how that religion differed from their religion, such as what holidays they celebrated and what rituals they practiced. The conversation consisted of statements such as, "We believe what Martin Luther said and you don't." "We believe in reincarnation, which is different than what you believe." There was no animosity; it was just a bunch of young children playing back what they had been told at home and sorting through the differences. This was new to me. I was raised in a nonreligious home and had never before heard descriptions of these religions or many of the words the other kids were saying. The conversation focused on the differences in their beliefs. I found this unsettling. If they believed different things, then shouldn't we all be trying to figure out which beliefs were right?

As I listened to the other kids talk about the differences in what they believed, I knew that they couldn't all be right. Even at that young age I had a deep sense that something was wrong. After everyone else had spoken, I was asked what my religion was. I answered that I wasn't sure, but I didn't think I had a religion. This created quite a stir, with several kids stating it wasn't possible. Finally, one kid asked, "Then what do you believe? You have to believe in something."

That playground conversation made a profound impression on me; I have thought about it many times since. What I found unsettling was not what they believed—rather, it was that the kids were willing to accept conflicting beliefs and not be bothered by it. It was as if we were all looking at a tree and one kid said, "My family believes that is an oak tree," another said, "My family believes it is a palm tree," and yet another said, "My family believes it isn't a tree. It's a tulip"—and yet no one was inclined to debate what the correct answer was.

Today I have a good understanding of how the brain forms beliefs. In the previous chapter, I described how the brain's model of the world can be inaccurate and why false beliefs can persist despite contrary evidence. For review, here are the three basic ingredients:

1. **Cannot directly experience**: False beliefs are almost always about things that we can't directly experience. If we cannot observe something directly—if we can't hear, touch, or see it ourselves—then we have to rely on what other people tell us. Who we listen to determines what we believe.

2. **Ignore contrary evidence**: To maintain a false belief, you have to dismiss evidence that contradicts it. Most false beliefs dictate behaviors and rationales for ignoring contrary evidence.

3. **Viral spread**: Viral false beliefs prescribe behaviors that encourage spreading the belief to other people.

Let's see how these attributes apply to three common beliefs that are almost certainly false.

Belief: Vaccinations Cause Autism

1. **Cannot directly experience**: No individual can directly sense whether vaccines cause autism; this requires a controlled study with many participants.
2. **Ignore contrary evidence**: You have to ignore the opinion of hundreds of scientists and medical personnel. Your rationale might be that these people are hiding the facts for personal gain or that they are ignorant of the truth.
3. **Viral spread**: You are told that by spreading this belief you are saving children from a debilitating condition. Therefore, you have a moral obligation to convince other people about the danger of vaccines.

Believing that vaccinations cause autism, even if it leads to the death of children, is not an existential threat to humanity. However, two common false beliefs that are existential threats are denying the danger of climate change and belief in an afterlife.

Belief: Climate Change Is Not a Threat

1. **Cannot directly experience**: Global climate change is not something individual people can observe. Your local weather has always been variable, and there have always been extreme weather events. Looking out your window day to day, you cannot detect climate change.
2. **Ignore contrary evidence**: Policies to fight climate change harm the near-term interests of some people and their businesses. Multiple rationales are used to protect these interests, such as that climate scientists are making up data and creating scary scenarios just to get more funding, or that scientific studies are flawed.

3. **Viral spread**: Climate-change deniers claim that policies to mitigate climate change are an attempt to take away personal freedoms, perhaps to form a global government or benefit a political party. Therefore, to protect freedom and liberty, you have a moral obligation to convince others that climate change is not a threat.

Hopefully it is obvious why climate change represents an existential risk to humanity. There is the possibility that we could alter Earth so much that it becomes uninhabitable. We don't know how likely this is, but we do know that our nearest neighboring planet, Mars, was once much more like Earth and is now an unlivable desert. Even if the possibility that this could happen to Earth is small, we need to be concerned.

Belief: There Is an Afterlife

Belief in an afterlife has been around for a very long time. It seems to occupy a persistent niche in the world of false beliefs.

1. **Cannot directly experience**: No one can directly observe the afterlife. It is by nature unobservable.
2. **Ignore contrary evidence**: Unlike the other false beliefs, there are no scientific studies that show it isn't true. Arguments against the existence of an afterlife are based mostly on lack of evidence. This makes it easier for believers to ignore claims that it does not exist.
3. **Viral spread**: Belief in the afterlife is viral. For example, a belief in heaven says your chance of going to heaven will increase if you try to convince others to believe in it as well.

Belief in an afterlife, on its own, is benign. For example, belief in reincarnation provides an incentive to live a more considerate life and seems to pose no existential risks. The threat arises if you believe that the afterlife is more important than the present life. At its extremes, this leads to the belief that destroying the Earth, or just several major cities and billions of people, will help you and your fellow believers achieve the desired afterlife. In the past, this might have led to the destruction and burning of a city or two. Today, it could lead to an escalating nuclear war that could make Earth unlivable.

The Big Idea

This chapter is not a comprehensive list of the threats we face, and I did not explore the full complexity of the threats that I did mention. The point I want to make is that our intelligence, which has led to our success as a species, could also be the seed of our demise. The structure of our brain, composed of an old brain and the neocortex, is the problem.

Our old brain is highly adapted for short-term survival and for having as many offspring as possible. The old brain has its good side, such as nurturing our young and caring for friends and relatives. But it also has its bad side, such as antisocial behavior to garner resources and reproductive access, including murder and rape. Calling these "good" and "bad" is somewhat subjective. From a replicating gene's point of view, they are all successful.

Our neocortex evolved to serve the old brain. The neocortex learns a model of the world that the old brain can use to better achieve its goals of survival and procreation. Somewhere along the evolutionary path, the neocortex gained the mechanisms for speech and high manual dexterity.

Language enabled the sharing of knowledge. This of course had huge advantages for survival, but it also sowed the seeds of false beliefs. Until the advent of language, the brain's model of the world was limited to only what we personally could observe. Language allowed us to expand our model to include things we learn from others. For example, a traveler might tell me that there are dangerous animals on the far side of a mountain—a place I have never been—and therefore extend my model of the world. However, the traveler's story might be false. Perhaps there are valuable resources on the far side of the mountain that the traveler does not want me to know about. In addition to language, our superior manual dexterity enabled the creation of sophisticated tools that include globe-spanning technologies that we increasingly rely on to support the large human population.

Now we find ourselves facing several existential threats. The first problem is that our old brain is still in charge and prevents us from making choices that support our long-term survival, such as reducing our population or eliminating nuclear weapons. The second problem is that the global technologies we have created are vulnerable to abuse by people with false beliefs. Just a few people with false beliefs can disrupt or misuse these technologies, such as by activating nuclear weapons. These people may believe that their actions are righteous and that they will be rewarded, perhaps in another life. Yet, the reality is that no such rewards would occur, while billions of people would suffer.

The neocortex has enabled us to become a technological species. We are able to control nature in ways that were unimaginable just one hundred years ago. Yet we are still a biological species. Each of us has an old brain that causes us to behave in ways that are detrimental to our species's long-term survival. Are we doomed? Is there any way out of this dilemma? In the remaining chapters, I describe our options.

CHAPTER 14

Merging Brains and Machines

There are two widely discussed proposals for how humans could combine brains and computers to prevent our death and extinction. One is *uploading* our brains into computers, and the other is *merging* our brains with computers. These proposals have been staples of science fiction and futurists for decades, but recently scientists and technologists have been taking them more seriously, and some people are working to make them a reality. In this chapter, I will explore these two proposals in light of what we have learned about brains.

Uploading your brain entails recording all the details of your brain, and then using them to simulate your brain on a computer. The simulator would be identical to your brain, so "you" would then live in the computer. The goal is to separate your mental and intellectual "you" from your biological body. This way, you can live indefinitely, including in a computer that is remote from Earth. You wouldn't die if Earth became uninhabitable.

Merging your brain with a computer entails connecting the neurons in your brain to the silicon chips in a computer. This would, for example, allow you to access all the resources of the internet just by thinking. One goal of this is to give you superhuman powers. Another is to mitigate the negative effects of an intelligence explosion, which (as I discussed in Chapter 11) is if intelligent machines suddenly get so smart that we can't control them, and then they kill or subjugate us. By merging our brains with computers, we also become superintelligent and are not left behind. We save ourselves by merging with the machines.

These ideas may strike you as ridiculous, outside the realm of possibility. But plenty of smart people take them seriously. It is easy to understand why they are appealing. Uploading your brain allows you to live forever, and merging your brain gives you superhuman abilities.

Will these proposals come to fruition, and will they mitigate the existential risks we face? I am not optimistic.

Why We Feel Trapped in Our Body

At times, I feel as if I am trapped in my body—as if my conscious intellect could exist in another form. Therefore, just because my body gets old and dies, why must "I" die? If I wasn't stuck in a biological body, couldn't I live forever?

Death is odd. On the one hand, our old brain is programmed to fear it, yet our bodies are programmed to die. Why would evolution make us fear the one thing that is most inevitable? Evolution settled on this conflicted strategy presumably for a good reason. My best guess is again based on the idea proposed by Richard Dawkins in his book *The Selfish Gene*. Dawkins argues that evolution is not about the survival of species, but about the survival of individual genes. From a gene's perspective, we need to live long enough to have children—that is, to make copies of the gene. Living much longer than that, although good for an individual animal, might not be in the best interest of an individual gene. For example, you and I are a particular combination of genes. After we have children, it might be better from a gene's perspective to make room for new combinations, new people. In a world with limited resources, it is better for a gene to exist in many different combinations with other genes, so that's why we are programmed to die—to make room for other combinations—but only after we have had offspring. The implication of Dawkins's theory is that we are unwitting servants to genes. Complex animals, such as ourselves, exist solely to help genes replicate. It is all about the gene.

Recently, something new has happened. Our species became intelligent. This of course helps us make more copies of our genes. Our intelligence lets us better avoid predators, find food, and live in varied ecosystems. But our emergent intelligence has had a consequence that is not necessarily in the best interest of genes. For the first time in the history of life on Earth, we understand what is going on. We have become enlightened. Our neocortex contains a model of evolution and a model of the universe and now it understands the truth underlying our existence. Because of our knowledge and intelligence, we can consider acting in ways that are not in the best interest of genes, such as using birth control or modifying genes that we don't like.

I see the current human situation as a battle between two powerful forces. In one corner, we have genes and evolution, which have dominated life for billions of years. Genes don't care about the survival of individuals. They don't care about the survival of our society. Most don't even care if our species goes extinct, because genes typically exist in multiple species. Genes only care about making copies of themselves. Of course, genes are just molecules and don't "care" about anything. But it useful to refer to them in anthropomorphic terms.

In the other corner, competing with our genes, is our newly emerged intelligence. The mental "I" that exists in our brains wants to break free from its genetic servitude, to no longer be held captive by the Darwinian processes that got us all here. We, as intelligent individuals, want to live forever and to preserve our society. We want to escape from the evolutionary forces that created us.

Uploading Your Brain

Uploading the brain into a computer is one means of escape. It would allow us to avoid the messiness of biology and live forever as a computer-simulated version of our former self. I wouldn't call brain uploading a mainstream idea, but it has been around for a long time and many people find it enticing.

Today, we don't have the knowledge or technology needed to upload a brain, but could we in the future? From a theoretical point of view, I don't see why not. However, it is so technically difficult that we may never be able to do it. But, regardless of whether it is technically feasible, I don't think it would be satisfactory. That is, even if you could upload your brain into a computer, I don't think you would like the result.

First let's discuss the feasibility of uploading our brains. The basic idea is that we make a map of every neuron and every synapse and then recreate all of this structure in software. The computer then simulates your brain, and, when it does, it will feel like you. "You" will be alive, but "you" will be in a computer brain instead of your old biological brain.

How much of your brain do we need to upload in order to upload you? The neocortex is obviously needed because it is the organ of thought and intelligence. Many of our day-to-day memories are formed in the hippocampal complex, so we need that too. What about all the emotional centers of the old brain? What about the brain stem and spinal cord? Our computer body wouldn't have lungs or a heart, so do we need to upload the parts of the brain that control them? Should we allow our uploaded brain to feel pain? You might think, "Of course not. We only want the good stuff!" But all the parts of our brain are interconnected in complex ways. If we didn't include everything, then the uploaded brain would have serious problems. Recall how a person can feel debilitating pain in a phantom limb, pain that results from a single missing limb. If we upload the neocortex, then it would have representations of every part of your body. If the body wasn't there, you might have severe pain everywhere. Similar problems would exist for every other part of the brain; if something is left out, then the other parts of the brain would be confused and not work properly. The fact is, if we want to upload you, and we want the uploaded brain to be normal, then we have to upload the entire brain, everything.

What about your body? You might think, "I don't need a body. As long as I can think and discuss ideas with other people, I will be happy." But your biological brain is designed to speak using your lungs and larynx, with their particular musculature, and your biological brain learned to see with your eyes, with their particular arrangement of photoreceptors. If your simulated brain is going to pick up thinking where your biological brain left off, then we need to recreate *your* eyes: eye muscles, retinas, etc. Of course, the uploaded brain doesn't need a physical body or physical eyes—a simulation should be sufficient. But it means that we would have to simulate your particular body and sensory organs. The brain and body are intimately wired together and, in many ways, are a singular system. We can't eliminate parts of the brain or parts of the body without seriously messing something up. None of this is a fundamental roadblock; it just means that it is far more difficult to upload you into a computer than most people imagine.

The next question we have to answer is how to "read" the details of your biological brain. How can we detect and measure everything in sufficient detail to recreate you in a computer? The human brain has about one hundred billion neurons and several hundred trillion synapses. Each neuron and synapse has a complex shape and internal structure. To recreate the brain in a computer, we have to take a snapshot that contains the location and structure of every neuron and every synapse. Today, we don't have the technology to do this in a dead brain, let alone a living one. Just the volume of data required to represent a brain greatly exceeds the capacity of our current computer systems. Obtaining the details needed to recreate you in a computer is so difficult that we might never be able to do it.

But let's put all these concerns aside. Let's say that sometime in the future we have the ability to instantaneously read out everything we need to recreate you in a computer. Let's assume we have computers with sufficient power to simulate you and your body. If we could do this, I have no doubts that the computer-based brain would be aware and conscious, just like you. But would you want this? Perhaps you are imagining one of the following scenarios.

You are at the end of your life. The doctor says you have just hours to live. At that moment, you flip a switch. Your mind goes blank. A few minutes later you wake up and find yourself living in a new computer-based body. Your memories are intact, you feel healthy again, and you begin your new eternal life. You shout, "Yay! I'm alive!"

Now imagine a slightly different scenario. Let's say we have the technology to read out your biological brain without affecting it. Now when you flip the switch, your brain is copied to a computer, but you feel nothing. Moments later, the computer says, "Yay! I'm alive." But you, the biological-you, are still here too. There are two of you now, one in a biological body and one in a computer body. The computer-you says, "Now that I'm uploaded, I don't need my old body, so please dispose of it." The biological-you says, "Wait a second. I'm still here, I don't feel any different, and I don't want to die." What should we do about this?

Maybe the solution to this dilemma is just to let the biological-you live out the rest of its life and die of natural causes. That seems fair. However, until that happens, there are two of you. The biological-you and the computer-you have different experiences. So, as time progresses, they grow apart and develop into different people. For example, the biological-you and the computer-you might develop different moral and political positions. Biological-you might regret creating computer-you. Computer-you might dislike having some old bio person claiming to be it.

Making matters worse, there would likely be pressure to upload your brain early in your life. For example, imagine that the intellectual health of computer-you depends on the intellectual health of biological-you at the time of upload. Therefore, to maximize the quality of life of your immortal copy, you should upload your brain when you are at maximum mental health, say at age thirty-five. Another reason you might want to upload your brain early in life is that every day you live in a biological body is a day that you might die by accident, and therefore lose the opportunity for immortality. So, you decide to upload yourself at age thirty-five. Ask yourself, would you (biological-you) at thirty-five years old feel comfortable killing yourself after making a copy of your brain? Would you (biological-you) even feel that you had achieved immortality as your computer copy went off on its own life and you slowly aged and died? I don't think so. "Uploading your brain" is a misleading phrase. What you have really done is split yourself into two people.

Now imagine that you upload your brain, and then the computer-you immediately makes three copies of itself. There are now four computer-yous and one biological-you. The five of you start having different experiences and drift apart. Each will be independently conscious. Have you become immortal? Which of the four computer-yous is your immortal you? As the biological-you slowly ages, moving toward death, it watches the four computer-yous go off to live their separate lives. There is no communal "you," just five individuals. They may have started with the same brain and memories, but they immediately become separate beings and thereafter live separate lives.

Perhaps you have noticed that these scenarios are just like having children. The big difference, of course, is that you don't upload your brain into your children's heads at birth. In some ways we try to do this. We tell our children about their family history and we train them to share our ethics and beliefs. In that way, we transfer some of our knowledge into our children's brains. But as they grow older, they have their own experiences and become separate people, just like an uploaded brain would do. Imagine if you could upload your brain into your children. Would you want to do it? If you did, I believe you would regret it. Your children would be saddled with memories of your past and would spend their lives trying to forget all the things you did.

Uploading your brain at first sounds like a great idea. Who wouldn't want to live forever? But making a copy of ourselves by uploading our brain into a computer will not achieve immortality any more than having children will. Copying yourself is a fork in the road, not an extension of it. Two sentient beings continue after the fork, not one. Once you realize this, then the appeal of uploading your brain begins to fade.

Merging Your Brain with a Computer

An alternative to uploading your brain is to merge it with a computer. In this scenario, electrodes are placed in your brain that are then connected to a computer. Now your brain can directly receive information from the computer, and the computer can directly receive information from your brain.

There are good reasons to connect brains to computers. For example, spinal-cord injuries can leave people with little or no ability to move. By implanting electrodes into the injured person's brain, the person can learn to control a robot arm or a computer mouse by thinking. Significant progress has already been achieved in this type of brain-controlled prosthetic, and it promises to improve the lives of many people. It does not take many connections from the brain to control a robot arm. For example, a few hundred or even a few dozen electrodes from the brain to a computer can be sufficient to control basic movements of a limb.

But some people dream of a deeper, more fully connected brain-machine interface, one where there are millions, perhaps billions, of connections going both ways. They hope this will give us amazing new abilities, such as accessing all the information on the internet as simply as we access our own memories. We could perform superfast calculations and data searches. We would thus radically enhance our mental abilities, merging brain with machine.

Similar to the "upload your brain" scenario, there are extreme technical challenges that have to be overcome to merge with a computer. These include how to implant millions of electrodes into a brain with minimal surgery, how to avoid rejection of the electrodes by our biological tissue, and how to reliably target millions of individual neurons. There are currently teams of engineers and scientists working on these problems. Once again, I don't want to focus on the technical challenges as much as the motivations and results. So, let's assume we can solve the technical problems. Why would we want to do this? Again, brain-computer interfaces make a lot of sense to help people with injuries. But why would we do this for healthy people?

As I mentioned, one prominent argument for merging your brain with a computer is to counter the threat of superintelligent AIs. Recall the intelligence explosion threat, where intelligent machines rapidly surpass us. I argued earlier that the intelligence explosion will not happen and is not an existential threat, but there are plenty of people who believe otherwise. They hope that by merging our brains with superintelligent computers, we would become superintelligent as well, and thus avoid being left behind. We are definitely entering science-fiction territory, but is it nonsense? I don't dismiss the idea of brain-computer interfaces for brain enhancement. The basic science needs to be pursued to restore movement to the injured. Along the way, we might discover other uses of the resulting technology.

For example, imagine we develop a way to precisely stimulate millions of individual neurons in the neocortex. Perhaps we do this by labeling individual neurons with barcode-like DNA snippets introduced via a virus (this kind of technology exists today). Then we activate these neurons using radio waves that are addressed to an individual cell's code (this technology does not exist, but it isn't out of the realm of possibility). Now we have a way of precisely controlling millions of neurons without surgery or implants. This could be used to restore sight to someone whose eyes don't function, or to create a new type of sensor, such as allowing someone to see using ultraviolet light. I doubt that we will ever completely merge our brain with a computer, but gaining new abilities is within the realm of likely advances.

In my opinion, the "uploading your brain" proposal offers few benefits and is so difficult that it is unlikely to ever happen. The "merging your brain with a computer" proposal will likely be achieved for limited purposes, but not to the point of fully uniting brain and machine. And a brain merged with a computer still has a biological brain and body that will decay and die.

Importantly, neither proposal addresses the existential risks facing humanity. If our species cannot live forever, are there things we can do today that would make our present existence meaningful, even when we are gone?

CHAPTER 15

Estate Planning for Humanity

Up until now, I have been discussing intelligence in both biological and machine form. From here on I want to shift the focus to knowledge. Knowledge is just the name for what we have learned about the world. Your knowledge is the model of the world that resides in your neocortex. Humanity's knowledge is the sum of what we have learned individually. In this and in the final chapter, I explore the idea that knowledge is worthy of preservation and propagation, even if that means doing so independently of humans.

I often think about dinosaurs. Dinosaurs lived on Earth for about 160 million years. They fought for food and territory and they struggled to not be eaten. Like us, they cared for their young and tried to protect their offspring from predators. They lived for tens of millions of generations, and now they are gone. What were their countless lives for? Did their onetime existence serve any purpose? Some dinosaur species evolved into today's birds, but most went extinct. If humans had not discovered dinosaur remains, then it is likely that nothing in the universe would ever know that dinosaurs existed.

Humans could suffer a similar fate. If our species becomes extinct, will anyone know that we once existed, that we once lived here on Earth? If no one finds our remains, then everything we have accomplished—our science, our arts, our culture, our history—would be lost forever. And being lost forever is the same as never existing. I find this possibility a bit unsatisfactory.

Of course, there are many ways that our individual lives can have meaning and purpose in the short term, in the here and now. We improve our communities. We raise and educate our children. We create works of art and enjoy nature. These types of activities can lead to a happy and fulfilling life. But these are personal and ephemeral benefits. They are meaningful to us while we and our loved ones are here, but any meaning or purpose diminishes over time, and it vanishes completely if our entire species goes extinct and no records remain.

It is almost a certainty that we, Homo sapiens, will become extinct at some point in the future. In several billion years, the Sun will die, ending life in our solar system. Before that happens, in a few hundred million to a billion years, the Sun will get hotter and greatly expand in size, making Earth a barren oven. These events are so far off that we don't need to worry about them now. But a much earlier extinction is possible. For example, the Earth could be hit by a large asteroid—unlikely in the short term, but it could happen at any time.

The most likely extinction risks we face in the short term—say, the next hundred or thousand years—are threats that we create. Many of our most powerful technologies have existed for only about one hundred years, and in that time we have created two existential threats: nuclear weapons and climate change. We will almost certainly create new threats as our technology advances. For example, we have recently learned how to precisely modify DNA. We might create new strains of viruses or bacteria that could literally kill every human. No one knows what will happen, but it is unlikely that we are done creating ways to destroy ourselves.

Of course, we need to do everything we can to mitigate these risks, and I am generally optimistic that we can prevent killing ourselves anytime soon. But I think it is a good idea to discuss what we can do now, just in case things don't work out so well.

Estate planning is something you do during your life that benefits the future, not yourself. Many people don't bother to do estate planning because they think there is nothing in it for them. But that isn't necessarily true. People who create estate plans often feel it provides a sense of purpose or creates a legacy. Plus, the process of establishing an estate plan forces you to think about life from a broad perspective. The time to do it is before you are on your deathbed, because by then you may no longer have the ability to plan and execute. The same holds true for estate planning for humanity. Now is a good time to think about the future and how we can influence it when we are no longer here.

When it comes to estate planning for humanity, who might benefit? Certainly no human, because the premise is that we are gone. The beneficiaries of our planning are other intelligent beings. Only an intelligent animal or an intelligent machine can appreciate our existence, our history, and our accumulated knowledge. I see two broad classes of future beings to think about. If humans go extinct but other life continues, then it is possible that intelligent animals might evolve a second time on Earth. Any second-species intelligent animal would certainly be interested in knowing as much as possible about the once-existing humans. We might call this the *Planet of the Apes* scenario, after the book and movie based on this premise. The second group we could try to reach are extraterrestrial intelligent species living elsewhere in our galaxy. The time of their existence might overlap with ours, or it might be far in the future. I will discuss both of these scenarios, although I believe that focusing on the latter is likely to be the most meaningful to us in the short term.

Why might other intelligent beings care about us? What can we do now that they would appreciate after we are gone? The most important thing is to let them know that we once existed. That fact alone is valuable. Think of how much we would appreciate knowing that intelligent life existed elsewhere in our galaxy. For many people, it would completely change their outlook on life. Even if we couldn't communicate with the extraterrestrial beings, it would be of immense interest to us to know that they exist or once existed. This is the goal of the search for extraterrestrial intelligence (SETI), a research program designed to find evidence of intelligent life in other parts of the galaxy.

Beyond the fact that we once existed, we could communicate our history and our knowledge. Imagine if the dinosaurs could tell us how they lived and what led to their demise. That would be immensely interesting and perhaps vitally useful to us. But because we are intelligent, we can tell the future far more valuable things than the dinosaurs could tell us. We have the potential to transfer all that we have learned. We might have scientific and technological knowledge that is more advanced than what the recipient knows. (Keep in mind that we are talking about what we know in the future, which will be more advanced than what we know now.) Again, think of how valuable it would be to us today if we could learn, for example, whether time travel is possible, or how to make a practical fusion reactor, or just what the answers are to fundamental questions, such as whether the universe is finite or infinite.

Finally, we might have the opportunity to convey what led to our demise. For example, if we, today, could learn that intelligent beings on distant planets went extinct due to self-induced climate change, we would take our current climate situation more seriously. Knowing how long other intelligent species existed and what led to their demise would help us to survive longer. It is hard to put a value on this type of knowledge.

I am going to discuss these ideas further by describing three scenarios we might use to communicate with the future.

Message in a Bottle

If you were stranded on a deserted island, you might write a message, put it in a bottle, and cast it into the sea. What would you write? You might put down where you are and hope that someone quickly discovers the message and rescues you, but you wouldn't have much hope that that would happen. It is more likely that your message is found long after you are gone. So instead, you might write about who you are and how you came to be stranded on the island. Your hope would be to have your fate known and recited by someone in the future. The bottle and your message are a means of not being forgotten.

The *Pioneer* planetary probes launched in the early 1970s have exited our solar system, passing into the great sea of space. Astronomer Carl Sagan advocated for including a plaque on the *Pioneer* probes. The plaques show where the craft came from and include a picture of a man and a woman. Later that decade, the *Voyager* probes similarly included a gold record that contained sounds and images from Earth. They, too, have left the solar system. We don't expect to ever see these spacecraft again. At the rate they are traveling, it will be tens of thousands of years before they could possibly reach another star. Although the probes were not designed for the purpose of communicating with distant aliens, they are our first messages in a bottle. They are mostly symbolic; not because of the time they will take to reach a potential audience, but because they will likely never be found. Space is so large, and the craft are so small, that the chance they encounter anything is tiny. Still, it is comforting to know that these spacecraft exist, traveling through space right now. If our solar system blew up tomorrow, these plaques and records would be the only physical record of life on Earth. They would be our only legacy.

Today, there are initiatives to send spacecraft to nearby stars. One prominent effort is called the Breakthrough Starshot. It envisions using high-powered, space-based lasers to propel tiny spacecraft to our nearest neighboring star, Alpha Centauri. The primary goal of this initiative is to take pictures of the planets orbiting Alpha Centauri and beam them back to Earth. Under optimistic assumptions, the entire process would take several decades.

Like the *Pioneer* and *Voyager* probes, the Starshot craft will continue to move through space long after we are gone. If the spacecraft are discovered by intelligent beings elsewhere in the galaxy, then those beings would know that we once existed and were intelligent enough to send spacecraft between the stars. Unfortunately, this is a poor way to intentionally communicate our existence to other beings. The spacecraft are tiny and slow. They can only reach a minuscule portion of our galaxy, and even if they reached an inhabited star system, the likelihood of being discovered is small.

Leave the Lights On

The SETI Institute has spent years trying to detect intelligent life elsewhere in our galaxy. SETI assumes that other intelligent beings are broadcasting a signal with sufficient power that we can detect it here on Earth. Our radar, radio, and TV broadcasts also send signals out into space, but these signals are so weak that we, using our existing SETI technology, couldn't detect similar signals from other planets unless they came from one very close to us. Hence, right now, there could be millions of planets with intelligent life just like ours, scattered throughout our galaxy, and—if each planet had a SETI program just like ours—nobody would detect anything. They, like us, would be saying, "Where is everybody?"

For SETI to be successful, we assume that intelligent beings are purposely creating strong signals designed to be detected over long distances. It is also possible that we could detect a signal not intended for us. That is, we might just happen to be aligned with a highly targeted signal and unintentionally pick up a conversation. But, for the most part, SETI assumes that an intelligent species is trying to make itself known by sending a powerful signal.

It would be considerate if we did the same. This is referred to as METI, which stands for messaging extraterrestrial intelligence. You might be surprised to learn that quite a few people think that METI is a bad idea—like, the worst idea ever. They fear that by sending a signal into space and thereby making our presence known, other, more advanced beings will come over to our star and kill us, enslave us, do experiments on us, or maybe accidentally infect us with a bug that we can't fight. Perhaps they are looking for planets that they can live on and the easiest way to find one is to wait for people like us to raise our hands and say, "Over here." In any case, humanity would be doomed.

This reminds me of one of the most common errors that first-time technology entrepreneurs make. They fear someone will steal their idea and therefore they keep it secret. In almost all cases, it is better to share your idea with anyone who might be able to help you. Other people can give you product and business advice and help you in numerous other ways. Entrepreneurs are far more likely to succeed by telling people what they are doing than by being secretive. It is human nature—aka old brain—to suspect everyone wants to steal your idea, where the reality is that you are lucky if anyone cares about your idea at all.

The risk of METI is built upon a series of improbable assumptions. It assumes that other intelligent beings are capable of interstellar travel. It assumes they are willing to spend significant time and energy to make the trip to Earth. Unless the aliens are hiding someplace nearby, it might take thousands of years for them to get here. It assumes that the intelligent agents need Earth or something on Earth that they can't get any other way, so it is worth the trip. It assumes that, despite having the technology for interstellar travel, they don't have the technology to detect life on Earth without us broadcasting our presence. And finally, it assumes that such an advanced civilization would be willing to cause us harm, as opposed to trying to help us or at least not hurt us.

Regarding this last point, it is a reasonable assumption that intelligent beings elsewhere in the galaxy evolved from nonintelligent life, just as we did. Therefore, the aliens probably faced the same types of existential risks that we face today. To survive long enough to become a galaxy-faring species means that they somehow got past these risks. Therefore, it is likely that whatever brain equivalent they have now would no longer be dominated by false beliefs or dangerously aggressive behavior. There is no guarantee this would happen, but it makes it less likely that they would harm us.

For all the above reasons, I believe we have nothing to fear from METI. Like a new entrepreneur, we will be better off trying to tell the world that we exist and hope that someone, anyone, cares.

The best way to approach both SETI and METI is critically dependent on how long intelligent life typically survives. It is possible that intelligence has arisen millions of times in our galaxy and almost none of the intelligent beings existed at the same time. Here is an analogy: Imagine fifty people are invited to an evening party. Everyone arrives at the party at a randomly chosen time. When they get there, they open the door and step inside. What are the chances they see a party going on or an empty room? It depends on how long they each stay. If all the partygoers stay for one minute before leaving, then almost everyone who shows up will see an empty room and conclude that no one else came to the party. If the partygoers stay for an hour or two, then the party will be a success, with lots of people in the room at the same time.

We don't know how long intelligent life typically lasts. The Milky Way galaxy is about thirteen billion years old. Let's say that it has been able to support intelligent life for about ten billion years. That is the length of our party. If we assume that humans survive as a technological species for ten thousand years, then it is as if we showed up for a six-hour party but only stayed for one fiftieth of a second. Even if tens of thousands of other intelligent beings show up for the same party, it is likely that we won't see anyone else while we are there. We will see an empty room. If we expect to discover intelligent life in our galaxy, it requires that intelligent life occurs often *and* that it lasts a long time.

I expect that extraterrestrial life is common. It is estimated there are about forty billion planets in the Milky Way alone that could support life, and life appeared on Earth billions of years ago, soon after the planet was formed. If Earth is typical, then life will be common in our galaxy.

I also believe that many planets with life will eventually evolve intelligent life. I have proposed that intelligence is based on brain mechanisms that first evolved for moving our bodies and recognizing places we have been. So, intelligence may not be too remarkable once there are multicellular animals moving about. However, we are interested in intelligent life that understands physics and that possesses the advanced technologies needed to send and receive signals from space. On Earth, this happened only once and only recently. We just don't have enough data to know how common species like us are. My guess is that technological species occur more often than you might conclude if you just looked at the history of Earth. I am surprised by how long it took for advanced technologies to appear on our planet. For example, I see no reason why technologically advanced species couldn't have appeared one hundred million years ago when dinosaurs roamed the Earth.

Regardless of how common technologically advanced life is, it might not last a long time. Technologically advanced species elsewhere in the galaxy will likely experience problems similar to those we face. The history of failed civilizations on Earth—plus the existential threats we are creating—suggests that advanced civilizations may not last very long. Of course, it is possible that species like ours could figure out how to survive for millions of years, but I don't consider it likely.

The implications of this is that intelligent and technologically advanced life might have sprung up millions of times in the Milky Way. But when we look out at the stars, we won't find intelligent life waiting to have a conversation with us. Instead, we will see stars where intelligent life once existed, but not now. The answer to the question "Where is everybody?" is that they already left the party.

There is a way to get around all these issues. There is a way to discover intelligent life in our galaxy, and perhaps even in other galaxies. Imagine that we create a signal that indicates we were here on Earth. The signal needs to be strong enough to be detected from far away, and it needs to last a long time. The signal needs to persist long after we are gone. Creating such a signal is like leaving a calling card at the party that says, "We were here." People who show up later won't find us, but they will know that we once existed.

This suggests a different way to think about SETI and METI. Specifically, it suggests we should first focus our efforts on how we could create a long-lasting signal. By long-lasting, I mean a hundred thousand years, or maybe millions of years or even a billion. The longer the signal persists, the more likely it will be successful. There is a nice secondary benefit to this idea: once we figure out how to make such a signal, then we will know what to look for ourselves. Other intelligent beings will likely reach the same conclusions we do. They, too, will look for how to create a long-lasting signal. As soon as we figure out how to do this, then we can start looking for it.

Today, SETI searches for radio signals that contain a pattern indicating that the signal was sent by an intelligent being. For example, a signal that repeated the first twenty digits of π would certainly be created by an intelligent species. I doubt we will ever find such a signal. It presumes that intelligent beings elsewhere in our galaxy have set up a powerful radio transmitter and, using computers and electronics, place a code in the signal. We have done this ourselves a few brief times. It requires a big antenna pointed into space, electrical energy, people, and computers. Because of the brief duration of the signals we sent, the messages were more symbolic efforts than serious attempts to reach out to the rest of our galaxy.

The problem with broadcasting a signal using electricity, computers, and antennas is that the system won't run for very long. Antennas, electronics, wires, etc. will not stay functional for even a hundred years without maintenance, let alone a million years. The method we choose to signal our presence has to be powerful, broadly directed, and self-sustaining. Once started, it needs to reliably run for millions of years without any maintenance or intervention. Stars are like this. Once started, a star emits large amounts of energy for billions of years. We want to find something that is similar but that couldn't be started without the guiding hand of an intelligent species.

Astronomers have found many odd sources of energy in the universe that, for example, oscillate, rotate, or emit short bursts. Astronomers look for natural explanations for these unusual signals, and usually they find them. Perhaps some of the as-yet-unexplained phenomena are not natural, but the type of signal I am talking about, created by intelligent beings. That would be nice, but I doubt it will be that easy. It is more likely that physicists and engineers will have to work on this problem for a while to come up with a set of possible methods to create a strong, self-sustaining signal that unmistakably originated with an intelligent being. The method also has to be something that we can implement. For example, physicists might conceive of a new type energy source capable of generating such a signal, but if we don't have the ability to create it ourselves, then we should assume that other intelligent beings can't either, and we should keep looking.

I have noodled on this problem for years, keeping an eye out for something that might fit the bill. Recently, a candidate surfaced. One of the most exciting areas of astronomy today is the discovery of planets orbiting other stars. Until recently, it was not known if planets were common or rare. We now know the answer: planets are common, and most stars have multiple planets, just like ours. The primary way we know this is by detecting slight reductions in starlight as a planet passes between a distant star and our telescopes. We could use the same basic idea to signal our presence. For example, imagine if we placed into orbit a set of objects that block a bit of the Sun's light in a pattern that would not occur naturally. These orbiting Sun blockers would continue to orbit the Sun for millions of years, long after we are gone, and they could be detected from far away.

We already possess the means to build such a Sun-blocker system, and there may be better ways to signal our presence. Here is not the place to evaluate our options. I am merely making the following observations: One, intelligent life may have evolved thousands or millions of times in our galaxy, but it is unlikely that we will find ourselves coexisting with other intelligent species. Two, SETI will be unlikely to succeed if we only look for signals that require ongoing participation by the sender. Three, METI is not only safe, it is the most important thing we can do to discover intelligent life in our galaxy. We need to first determine how we can make our presence known in a way that lasts for millions of years. Only then will we know what to look for.

Wiki Earth

Letting a distant civilization know that we once existed is an important first goal. But to me, the most important thing about humans is our knowledge. We are the only species on Earth that possesses knowledge of the universe and how it works. Knowledge is rare, and we should attempt to preserve it.

Let's assume that humans go extinct, but life on Earth continues. For example, an asteroid is believed to have killed the dinosaurs and many other species, but some small animals managed to survive the impact. Sixty million years later, some of those survivors became us. This actually happened and it could happen again. Imagine now that we humans have gone extinct, perhaps due to a natural disaster or to something we did. Other species survive, and in fifty million years one of them becomes intelligent. That species would definitely want to know everything they could about the long-gone human epoch. They would be particularly interested in knowing the extent of our knowledge, and what happened to us.

If humans go extinct, then in just a million years or so all detailed records of our life will likely be lost. There will be buried remains of some of our cities and large infrastructure, but pretty much every document, film, and recording will no longer exist. Future nonhuman archeologists will struggle to piece together our history in the same way paleontologists today struggle to figure out what happened to the dinosaurs.

As part of our estate plan, we could preserve our knowledge in a more permanent form, one that could last tens of millions of years. There are several ways we might do this. For example, we could continually archive a knowledge base such as Wikipedia. Wikipedia is itself constantly updated, so it would document events up to the point where our society started to fail, it covers a broad range of topics, and the archive process could be automated. The archive shouldn't be located on Earth, as Earth might be partially destroyed in a singular event, and over millions of years little will remain intact. To overcome this problem, we could locate our archive in a set of satellites that orbit the Sun. This way, the archive will be easy to discover but difficult to physically alter or destroy.

We would design the satellite-based archive so that we could send automatic updates to it but its content could not be erased. The electronics in the satellite will stop functioning shortly after we are gone, so, to read the archive, a future intelligent species would have to develop the technology to travel to the archive, bring it back to Earth, and extract the data. We could use multiple satellites at different orbits for redundancy. We already possess the ability to create a satellite archive and retrieve it. Imagine if a previous intelligent species on Earth had placed a set of satellites around the solar system. We would have discovered them by now and already brought them back to Earth.

In essence, we could create a time capsule designed to last for millions or hundreds of millions of years. In the distant future, intelligent beings—whether they evolve on Earth or travel from another star—could discover the time capsule and read its contents. We won't know whether our repository will be discovered or not; that's the nature of estate plans. If we do this, and it is read in the future, imagine how appreciative the recipients would be. All you have to do is think of how excited we would be to discover such a time capsule ourselves.

An estate plan for humanity is similar to an estate plan for individuals. We would like our species to live forever, and maybe that will happen. But it is prudent to put in place a plan just in case the miracle doesn't happen. I have suggested several ideas we could pursue. One is to archive our history and knowledge in a way that future intelligent species on Earth could learn about humanity— what we knew, our history, and what ultimately happened to us. Another is to create a long-lasting signal, one that tells intelligent beings elsewhere in space and time that intelligent humans once lived around the star we call the Sun. The beauty of the long-lasting signal is that it might help us in the short term by leading us to discover that other intelligent species preceded us.

Is it worth the time and money to pursue initiatives like this? Would it be better to put all our efforts into improving life on Earth? There is always a friction between investing in the short term and investing in the long term. Short-term problems are more pressing, whereas investing in the future has few immediate benefits. Every organization—whether it be a government, a business, or a household—faces this dilemma. However, not investing in the long term guarantees future failure. In this case, I believe that investing in an estate plan for humanity has several near-term benefits. It will keep us more aware of the existential threats we face. It will propel more people to think about the long-term consequences of our actions as a species. And it will provide a type of purpose to our life should we eventually fail.

CHAPTER 16

Genes Versus Knowledge

"Old Brain—New Brain" is the title of the first chapter in this book. It is also an underlying theme. Recall that 30 percent of our brain, the old brain, is composed of many different parts. These old-brain areas control our bodily functions, basic behaviors, and emotions. Some of these behaviors and emotions cause us to be aggressive, violent, and covetous, to lie and to cheat. Every one of us harbors these tendencies to one degree or another, because evolution discovered they are useful for propagating genes. Seventy percent of our brain, the new brain, is made of one thing: the neocortex. The neocortex learns a model of the world, and it is this model that makes us intelligent. Intelligence evolved because it, too, is useful for propagating genes. We are here as servants to our genes, but the balance of power between the old brain and new brain has started to shift.

For millions of years, our ancestors had limited knowledge of our planet and the broader universe. They only understood what they could personally experience. They didn't know the size of the Earth or that it was a sphere. They didn't know what the Sun, Moon, planets, and stars were and why they moved through the sky as they do. They had no idea how old the Earth is and how its varied life-forms came to be. Our ancestors were ignorant about the most basic facts of our existence. They made up stories about these mysteries, but the stories were not true.

Recently, using our intelligence, we have not only solved the mysteries that vexed our ancestors but the pace of scientific discovery is accelerating. We know how incredibly big the universe is and how incredibly small we are. We now understand that our planet is billions of years old and that life on Earth has been evolving also for billions of years. Fortunately, the entire universe appears to operate by one set of laws, some of which we have discovered. It appears tantalizingly possible that we might be able to discover all of them. Millions of people around the world are actively working on scientific discovery in general, and billions more feel connected to the mission. It is an incredibly exciting time to be alive.

However, we have a problem that could quickly halt our race to enlightenment, and might end our species entirely. Earlier, I explained that no matter how smart we become, our neocortex remains connected to the old brain. As our technologies become more and more powerful, the selfish and shortsighted behaviors of the old brain could lead us to extinction or plunge us into societal collapse and another dark age. Compounding this risk is that billions of humans still have false beliefs about the most fundamental aspects of life and the universe. Viral false beliefs are another source of behaviors that threaten our survival.

We face a dilemma. "We"—the intelligent model of ourselves residing in the neocortex—are trapped. We are trapped in a body that not only is programmed to die but is largely under the control of an ignorant brute, the old brain. We can use our intelligence to imagine a better future, and we can take actions to achieve the future we desire. But the old brain could ruin everything. It generates behaviors that have helped genes replicate in the past, yet many of those behaviors are not pretty. We try to control our old brain's destructive and divisive impulses, but so far we have not been able to do this entirely. Many countries on Earth are still ruled by autocrats and dictators whose motivations are largely driven by their old brain: wealth, sex, and alpha-male-type dominance. The populist movements that support autocrats are also based on old-brain traits such as racism and xenophobia.

What should we do about this? In the previous chapter, I discussed ways we could preserve our knowledge in case humanity doesn't survive. In this final chapter, I discuss three methods we might pursue to prevent our demise. The first method may or may not work without modifying our genes, the second is based on gene modification, and the third abandons biology altogether.

These ideas might strike you as extreme. However, ask yourself: What is the purpose of living? What are we trying to preserve when we struggle to survive? In the past, living was always about preserving and replicating genes, whether we realized it or not. But is that the best way forward? What if instead we decide that living should focus on intelligence and the preservation of knowledge. If we were to make that choice, then what we consider extreme today might be just the logical thing to do in the future. The three ideas that I present here are, in my opinion, possible and have a high likelihood of being pursued in the future. They may seem unlikely now, in the same way that handheld computers seemed unlikely in 1992. We will have to let time play out to see which, if any, turns out to be viable.

Become a Multi-planet Species

When our Sun dies, all life in our solar system will die too. But most extinction events that concern us would be localized to Earth. If a large asteroid hit Earth, for example, or we made it uninhabitable in a nuclear war, other nearby planets would be unaffected. Therefore, one way to lessen the risk of extinction is to become a two-planet species. If we could establish a permanent presence on another nearby planet or moon, then our species and our accumulated knowledge might survive even if Earth became uninhabitable. This logic is one of the driving forces behind the current efforts to put people on Mars, which seems to be the best option for locating a human colony. I find the possibility of traveling to other planets exciting. It has been a long time since we have traveled to new and unexplored destinations.

The main difficulty with living on Mars is that Mars is a terrible place to live. The lack of a significant atmosphere means that a short exposure outside will kill you, and a leak in your roof or a broken window could kill your entire family. Radiation from the Sun is higher on Mars and is also a major risk to living there, so you would have to continually protect yourself from the Sun. Mars soil is poisonous and there is no surface water. Seriously, it is easier to live on the South Pole than on Mars. But that doesn't mean we should give up on the idea. I believe we could live on Mars, but to do so we need something we don't yet have. We need intelligent and autonomous robots.

For humans to live on Mars, we would need large, airtight buildings to live and grow food in. We would need to extract water and minerals from mines and to manufacture air to breathe. Ultimately, we would need to terraform Mars to reintroduce an atmosphere. These are huge infrastructure projects that could take decades or centuries to complete. Until Mars became self-sufficient, we would have to send everything needed: food, air, water, medicine, tools, construction equipment, materials, and people—lots of people. All the work would have to be done wearing cumbersome space suits. It is hard to overstate the difficulty humans would face trying to construct livable environments and all the infrastructure needed to create a permanent self-sustaining Mars colony. The loss of life, the psychological damage, and the financial costs would be enormous, probably larger than we are willing to endure.

But preparing Mars for humans could be accomplished if, instead of sending human engineers and construction workers, we sent intelligent robotic engineers and construction workers. They would get their energy from the Sun and could work outside without needing food, water, or oxygen. They could work tirelessly with no emotional stress for as long as required to make Mars safe for human habitation. The robotic corps of engineers would need to work mostly autonomously. If they relied on constant communication with Earth, progress would be too slow.

I have never been a fan of science-fiction literature, and this scenario sounds suspiciously like science fiction. Yet I see no reason why we can't do this, and, if we want to become a multi-planet species, I believe we have no choice. For humans to live on Mars on a permanent basis requires intelligent machines to help us. The key requirement is to endow the Mars robotic workforce with the equivalent of a neocortex. The robots need to use complex tools, manipulate materials, solve unanticipated problems, and communicate with each other, similar to how people do. I believe the only way we can achieve this is to finish reverse engineering the neocortex and create equivalent structures in silicon. Autonomous robots need to have a brain built on the principles I outlined earlier, the principles of the Thousand Brains Theory of Intelligence.

Creating truly intelligent robots is achievable, and I am certain it will happen. I believe we could do this within several decades if we made it a priority. Fortunately, there are plenty of terrestrial reasons to build intelligent robots too. Therefore, even if we don't make it a national or an international priority, market forces will eventually fund the development of machine intelligence and robotics. I hope that people around the world will come to understand that being a multi-planet species is an exciting goal important for our survival, and that intelligent robotic construction workers are necessary to achieve it.

Even if we create intelligent robotic workers, terraform Mars, and establish colonies of humans, we will still have a problem. The humans who go to Mars will be just like the humans on Earth. They will have an old brain and all of the complications and risks that accompany it. Humans living on Mars will fight for territory, make decisions based on false beliefs, and probably create new existential risks for those living there.

History suggests that eventually the people living on Mars and the people living on Earth will end up feuding in ways that could endanger one or both populations. For example, imagine that two hundred years from now, ten million humans are living on Mars. But then, something goes bad on Earth. Perhaps we accidentally poison most of the planet with radioactive elements, or Earth's climate starts rapidly failing. What would happen? Billions of Earth inhabitants might suddenly want to move to Mars. If you let your imagination run a bit, you can see how this could easily turn out badly for everyone. I don't want to speculate on negative outcomes. But it is important to recognize that becoming a multi-planet species is no panacea. Humans are humans, and the problems we create on Earth will also exist on other planets we inhabit.

What about becoming a multi-stellar species? If humans could colonize other star systems, then we could expand throughout the galaxy and the chance that some of our descendants would survive indefinitely would increase dramatically.

Is interstellar human travel possible? On the one hand, it seems like it should be. There are four stars that are less than five light-years from us and eleven stars less than ten light-years away. Einstein showed that it is impossible to accelerate to the speed of light, so let's say we travel at half that speed. Then a mission to a nearby star could be accomplished in a decade or two. On the other hand, we don't know how to get anywhere close to this speed. Using the technologies we have today, it would take tens of thousands of years to reach our nearest neighboring star. Humans cannot make that long a trip.

There are many physicists thinking of clever ways to overcome the problems of interstellar flight. Perhaps they will discover ways to travel close to the speed of light, or even faster than it. Many things that seemed impossible just two hundred years ago are now commonplace. Imagine that you spoke at a meeting of scientists in 1820 and said that in the future anyone could travel in comfort from continent to continent in a matter of hours, or that people would routinely hold face-to-face conversations with other people anywhere in the world by looking at and speaking to their hand. Nobody would have thought these activities would *ever* be possible, but here we are. The future will certainly surprise us with new advances that today are inconceivable, and one of them may be practical space travel. But I feel comfortable in predicting that human interstellar travel will not happen in less than fifty years. I would not be surprised if it never happened.

I still advocate for becoming a multi-planet species. It will be an inspiring exploratory adventure, and it might reduce the near-term risk of humans going extinct. But the inherent risks and limitations that arise from our evolutionary heritage remain. Even if we manage to establish colonies on Mars, we may have to accept that we will never venture beyond our solar system.

However, we have other options. These require us to look at ourselves objectively and ask, What is it about humanity we are trying to preserve? I'll address that question first, before discussing two more options for securing our future.

Choosing Our Future

Starting with the Enlightenment at the end of the eighteenth century, we have accumulated increasing evidence that the universe progresses without a guiding hand. The emergence of simple life, then complex organisms, and then intelligence was neither planned nor inevitable. Similarly, the future of life on Earth and the future of intelligence are not predetermined. It appears that the only thing in the universe that cares how our future unfolds is us. The only desirable future is the one that *we* desire.

You might object to this statement. You might say that there are many other species living on Earth, some also intelligent. We have harmed many of these species and caused others to go extinct. Shouldn't we consider what other species "desire"? Yes, but it is not so simple.

Earth is dynamic. The tectonic plates that make its surface are constantly moving, creating new mountains, new continents, and new seas while plunging existing features into the Earth's center. Life is similarly dynamic. Species are ever changing. We are not the same, genetically, as our ancestors who lived one hundred thousand years ago. The rate of change may be slow, but it never stops. If you look at Earth this way, then it doesn't make sense to try to preserve species or to preserve Earth. We cannot stop the Earth's most basic geological features from changing, and we can't stop species from evolving and going extinct.

One of my favorite activities is wilderness hiking, and I consider myself to be an environmentalist. But I don't pretend that environmentalism is about preserving nature. Every environmentalist would be happy to see the extinction of some living things—say, the poliovirus—while simultaneously going to great lengths to save an endangered wildflower. From the universe's perspective, this is an arbitrary distinction; neither the poliovirus nor the wildflower is better or worse than the other. We make the choice about what to protect based on what is in our best interest.

Environmentalism is not about preserving nature, but about the choices we make. As a rule, environmentalists make choices that benefit future humans. We try to slow down changes to the things we like, such as wilderness areas, to increase the chances that our descendants can also enjoy these things. There are other people who would choose to turn wilderness areas into strip mines so that they can benefit today, more of an old-brain choice. The universe doesn't care which option we choose. It is our choice whether we help future humans or present humans.

There is no option to do nothing. As intelligent beings, we must make choices, and our choices will steer the future one way or another. As to the other animals on Earth, we can choose to help them or not. But as long as we are here, there is no option to let things go their "natural" way. We are part of nature and we must make choices that will impact the future.

As I see it, we have a profound choice to make. It is a choice between favoring the old brain or the new brain. More specifically, do we want our future to be driven by the processes that got us here, namely, natural selection, competition, and the drive of selfish genes? Or do we want our future to be driven by intelligence and the desire to understand the world? We have the opportunity to choose between a future where the primary driver is the creation and dissemination of knowledge and a future where the primary driver is the copying and dissemination of genes.

To exercise our choice, we need the ability to change the course of evolution by manipulating genes and the ability to create intelligence in nonbiological form. We already have the former, and the latter is imminent. Use of these technologies has led to ethical debates. Should we manipulate the genes of other species to improve our food supply? Should we manipulate our own genes to "improve" our progeny? Should we create intelligent machines that are smarter and more capable than we are?

Perhaps you have already formed opinions on these questions. You might think that these things are fine, or you might think that they are unethical. Regardless, I don't see any harm in discussing our options. Looking carefully at our choices will help us make informed decisions, no matter what we choose to do.

Becoming a multi-planet species is an attempt to prevent our extinction, but it is still a future dictated by genes. What kind of choices could we make to favor the propagation of knowledge over the propagation of genes?

Modify Our Genes

We have recently developed the technology to precisely edit DNA molecules. Soon, we will be able to create new genomes and modify existing ones with the precision and ease of creating and editing text documents. The benefits of gene editing could be huge. For example, we could eliminate inherited diseases that cause suffering in millions of people. However, the same technology can also be used to design entirely new life-forms or to modify our children's DNA—for example, to make them better athletes or more attractive. Whether we think this kind of manipulation is OK or abhorrent depends on the circumstances. Modifying our DNA to make us look more attractive seems unnecessary, but if gene editing keeps our entire species from going extinct then it becomes an imperative.

For example, let's say we decide that establishing a colony on Mars is a good insurance plan for the long-term survival of our species, and many people sign up to go. But then we discover that humans cannot live on Mars for extended periods of time due to the low gravity. We already know that spending months in zero gravity on the International Space Station leads to medical problems. Perhaps after ten years of living in the low gravity of Mars our bodies start failing and die. A permanent population on Mars would then seem impossible. However, let's say we could fix this problem by editing the human genome, and people with these DNA modifications could live indefinitely on Mars. Should we allow people to edit their genes and the genes of their children so that they can live on Mars? Anyone willing to go to Mars is already accepting life-threatening risks. And the genes of people living on Mars will slowly change anyway. So why shouldn't people be able to make that choice? If you think this form of gene editing should be forbidden, would you change your mind if Earth was becoming uninhabitable and the only way you could survive at all was to move to Mars?

Now imagine we learn how to modify our genes to eliminate aggressive behavior and make a person more altruistic. Should we allow this? Consider that when we select who gets to be an astronaut, we pick people who naturally have these attributes. There is a good reason we do this; it increases the likelihood that a space mission succeeds. If in the future we send people to live on Mars, we will likely do the same type of screening. Wouldn't we give preference to emotionally stable people over people with a short fuse and a history of aggression? When a single careless or violent act could kill an entire community, wouldn't the people already living on Mars demand that new arrivals pass some sort of emotional stability test? If we could make a better citizen by editing DNA, the existing Mars inhabitants might insist on that.

Consider one more hypothetical scenario. Some fish can survive being frozen in ice. What if we could modify our DNA so that a human could be similarly frozen and then thawed sometime in the future? I can imagine many people would want to freeze their body to be woken up again in one hundred years. It would be thrilling to live out the last ten or twenty years of your life in the future. Would we allow it? What if this modification allowed humans to travel to other stars? Even if the trip took thousands of years, our space travelers could be frozen on departure and thawed when they got to their destination. There would be no shortage of volunteers for such a trip. Are there reasons we should forbid the DNA modifications that would make such a voyage possible?

I can come up with many scenarios where we might decide it is in our best personal interest to significantly modify our DNA. There is no absolute right or wrong; there are only choices that we get to make. If people say we should never allow DNA editing on principle, then, whether they realize it or not, they have chosen a future that is in the best interests of our existing genes or, as is often the case, of viral false beliefs. By taking such a stand, they are eliminating choices that might be in the best interest of the long-term survival of humanity and the long-term survival of knowledge.

I am not advocating that we should edit the human genome with no oversight or deliberation. And nothing I have described involves coercion. No one should ever be forced to do any of these things. I am only pointing out that gene editing is possible, and therefore we have choices. Personally, I don't see why the path of unguided evolution is preferable over a path of our own choosing. We can be thankful that evolutionary processes got us here. But now that we are here, we have the option to use our intelligence to take control of the future. Our survival as a species and the survival of our knowledge might be more secure if we do.

A future designed by editing our DNA is still a biological future, and that imposes limits on what is possible. For example, it is unclear how much can be accomplished by DNA editing. Will it be possible to edit our genome to allow future humans to travel between the stars? Will it be possible to make future humans that don't kill each other on a remote planetary outpost? No one knows. Today, we don't have enough knowledge about DNA to predict what is possible and what is not. I would not be surprised if we found that some of the things we might want to do are impossible in principle.

Now I turn to our final option. It is perhaps the surest way to guarantee the preservation of knowledge and the survival of intelligence, but it might also be the most difficult.

Leaving Darwin's Orbit

The ultimate way to free our intelligence from the grip of our old brain and our biology is to create machines that are intelligent like us, but not dependent upon us. They would be intelligent agents that could travel beyond our solar system and survive longer than we will. These machines would share our knowledge but not our genes. If humans should regress culturally—as in a new dark age—or if we become extinct, our intelligent machine progeny would continue on without us.

I hesitate to use the word "machine," because it might invoke an image of something like a computer sitting on a desk, or a humanoid robot, or some evil character from a science-fiction story. As I described earlier, we cannot predict what intelligent machines will look like in the future, in the same way that the early designers of computers couldn't imagine what future computers would look like. No one in the 1940s imagined that computers could be smaller than a grain of rice, small enough to be embedded into nearly everything. Nor could they imagine powerful cloud computers that are accessible everywhere but not exactly located anywhere.

Similarly, we cannot imagine what intelligent machines of the future will look like, or what they will be made of, so let's not try. It might limit our thinking of what is possible. Instead, let's discuss the two reasons why we might want to create intelligent machines that can travel to the stars without us.

Goal Number One: Preserve Knowledge

In the previous chapter, I described how we might preserve knowledge in a repository orbiting the Sun. I labeled this Wiki Earth. The repository I described was static. It is like a library of printed books floating in space. Our goal in creating it would be to preserve knowledge, with the hope that some future intelligent agent discovers the repository and figures out how to read its contents. However, without humans actively looking after its maintenance, the repository will slowly decay. Wiki Earth does not make copies of itself, does not repair itself, and, therefore, is temporary. We would design it to last a long time, but at some point in the distant future it would cease to be readable.

The human neocortex is also like a library. It contains knowledge about the world. But unlike Wiki Earth, the neocortex makes copies of what it knows by transferring its knowledge to other humans. For example, this book is an attempt by me to transfer some things I know to other people, like you. This ensures that knowledge is distributed. The loss of any one person will not lead to the permanent loss of knowledge. The surest way to preserve knowledge is to continually make copies.

Therefore, one goal of creating intelligent machines would be to replicate what humans are already doing: preserving knowledge by making and distributing copies. We would want to use intelligent machines for this purpose because they could continue preserving knowledge long after we are gone, and they can distribute knowledge to places we can't go, such as other stars. Unlike humans, intelligent machines could slowly spread across the galaxy. They hopefully would be able to share knowledge with intelligent beings elsewhere in the universe. Imagine how exciting it would be if we discovered a repository of knowledge and galactic history that had traveled to our solar system.

In the previous chapter on estate planning, I described both the Wiki Earth idea and the idea of creating a long-lasting signal to indicate that we, an intelligent species, once existed in our solar system. Together, these two systems could potentially direct other intelligent beings to our solar system and then to the discovery of our knowledge repository. What I am proposing in this chapter is a different way of achieving a similar result. Instead of directing alien intelligence to the repository of knowledge in our solar system, we send copies of our knowledge and history throughout the galaxy. Either way, something that is intelligent must make the long trip through space.

Everything wears down. As intelligent machines travel through space, some will be damaged, lost, or unintentionally destroyed. Therefore, our intelligent machine progeny must be able to repair themselves, and, when needed, make copies of themselves. I realize this will scare people who worry about intelligent machines taking over the world. As I explained earlier, I don't believe we have to worry about this, as most intelligent machines won't be able to make copies of themselves. But in this scenario, it is a requirement. However, it is so hard for intelligent machines to replicate that it is the main reason this scenario might not be possible. Imagine a handful of intelligent machines traveling through space. After thousands of years, they arrive at a new solar system. They find mostly barren planets and one planet with primitive, single-celled life. This is what would have been found by a visitor to our solar system a few billion years ago. Now let's say the intelligent machines decide they need to replace two of their members and create a few new intelligent machines to send off to another star. How could they do this? If, for example, the machines were built using silicon chips like the ones we use in computers, then would they need to build silicon-chip-manufacturing plants and all the necessary supply chains? This might be infeasible. Perhaps we will learn how to create intelligent machines that are able to replicate using common elements, similar to carbon-based life on Earth.

I don't know how to overcome the many practical problems that interstellar travel presents. But, once again, I believe we shouldn't focus on the physical manifestations of future intelligent machines. There may be ways of building intelligent machines using materials and construction methods that we have not yet invented. For now, it is more important to discuss goals and concepts to help us decide if this is something we would choose to do if we could. If we decide that sending intelligent machines to explore the galaxy and spread knowledge is something we want to pursue, then it is possible we could devise ways of overcoming the obstacles.

Goal Number Two: Acquire New Knowledge

If we managed to create self-sustaining intelligent machines that traveled between stars, they would discover new things. They would undoubtedly discover new types of planets and stars, and make other discoveries that we can't imagine. Perhaps they would discover the answers to deep mysteries about the universe, such as its origin or destiny. That is the nature of exploration: you don't know what you will learn, but you will learn something. If we sent humans to explore the galaxy, we would expect them to make discoveries. In many ways, intelligent machines will be more capable of discovery than humans. Their brain equivalent will have more memory, work faster, and have novel sensors. They would be better scientists than we are. If intelligent machines traversed our galaxy, they would continually increase what is known about the universe.

A Future with Purpose and Direction

Humans have long dreamed about traveling between the stars. Why?

One reason is to extend and preserve our genes. This is based on the idea that the destiny of a species is to continually explore new lands and establish colonies wherever it can. We have done this repeatedly in the past, traveling over mountains and across oceans to establish new societies. This serves the interests of our genes, and therefore we are programmed to explore. Curiosity is one of our old-brain functions. It is hard to resist exploring, even when it would be safer not to. If humans could travel to the stars, it would just be an extension of what we have always done, spreading our genes to as many places as possible.

The second reason, the one that I have suggested in this chapter, is to extend and preserve our knowledge. This line of thinking is based on the assumption that intelligence, not our particular genes, is why our species is important. Therefore, we should travel to the stars to learn more and safeguard our knowledge for the future.

But is that a better choice? What is wrong with continuing on as we always have? We could forget all this chatter about preserving knowledge or creating intelligent machines. Life on Earth has been pretty good so far. If humans can't travel to other stars, so what? Why not just continue on as we have and enjoy the ride while it lasts?

That is a reasonable choice, and in the end, it may be our only one. But I want to make the case for knowledge over genes. There is a fundamental difference between the two, a difference that makes preserving and spreading knowledge, in my opinion, a more worthy goal than preserving and spreading our genes.

Genes are just molecules that replicate. As genes evolve, they are not heading in any particular direction, nor is one gene intrinsically better than another, just as one molecule is not intrinsically better than any other molecule. Some genes may be better at replication, yet, as environments change, which genes are better at replicating also changes. Importantly, there is no overall direction to the changes. Life based on genes has no direction or goal. Life may manifest itself as a virus, a single-celled bacterium, or a tree. But there doesn't appear to be any reason to suggest one life-form is better than another, beyond its ability to replicate.

Knowledge is different. Knowledge has both a direction and an end goal. For example, consider gravity. In the not-too-distant past, nobody had any idea why things fell down and not up. Newton created the first successful theory of gravity. He proposed that it is a universal force, and he showed that it behaves according to a set of simple laws that could be expressed mathematically. After Newton, we would never go back to having no theory of gravity. Einstein's explanation of gravity is better than Newton's, and we will never go back to Newton's theory. It wasn't that Newton was wrong. His equations still accurately describe gravity as we experience it every day. Einstein's theory incorporates Newton's but better describes gravity under unusual conditions. There is a direction to knowledge. Knowledge of gravity can go from no knowledge, to Newton's, to Einstein's, but it can't go in the opposite direction.

In addition to a direction, knowledge has an end goal. The earliest human explorers did not know how big the Earth was. No matter how far they traveled, there was always more. Was the Earth infinite? Did it end with an edge where further travel would cause you to fall off? Nobody knew. But there was an end goal. It was assumed that there was an answer to the question, How big is the Earth? We eventually achieved that goal with a surprising answer. The Earth is a sphere, and now we know how big the Earth is.

We are facing similar mysteries today. How big is the universe? Does it go on forever? Does it have an edge? Does it wrap around on itself like the Earth? Are there many universes? There are plenty of other things we don't understand: What is time? How did life originate? How common is intelligent life? Answering these questions is a goal, and history suggests we can achieve it.

A future driven by genes has little to no direction and only short-term goals: stay healthy, have kids, enjoy life. A future designed in the best interest of knowledge has both direction and end goals.

The good news is we don't have to choose one future over the other. It is possible to do both. We can continue to live on Earth, doing our best to keep it livable and trying to protect ourselves from our own worst behaviors. And we can simultaneously dedicate resources to ensuring the preservation of knowledge and the continuation of intelligence for a time in the future when we are no longer here.

I wrote Part 3 of this book, the last five chapters, to make a case for knowledge over genes. I asked you to look at humans objectively. I asked you to see how we make poor decisions and why our brains are susceptible to false beliefs. I asked you to consider knowledge and intelligence as more precious than genes and biology and, therefore, worthy of preservation beyond their current home in our biological brain. I asked you to consider the possibility of progeny based on intelligence and knowledge, and that these descendants might be equally worthy to ones based on genes.

I want to emphasize again that I am not prescribing what we should do. My goal is to encourage discussion, to point out that some things we think of as ethical certainties are actually choices, and to bring some underserved ideas to the forefront.

Now I want to return to the present.

Final Thoughts

I have a vision that never ceases to entertain me. I imagine the vast universe, with its hundreds of billions of galaxies. Each galaxy contains hundreds of billions of stars. Around each star, I picture planets of limitless variety. I imagine these trillions of monstrously sized objects slowly orbiting each other in the vast emptiness of space for billions of years. What amazes me is that the only thing in the universe that knows about this—the only thing that knows that the universe exists at all—is our brain. If it wasn't for brains, then nothing would know that anything exists. It prompts the question that I mentioned at the beginning of the book: If there is no knowledge of something, can we say that the thing exists at all? That our brain plays such a unique role is fascinating. Of course, there may be intelligent beings elsewhere in the universe, but this makes it even more entertaining to think about.

Thinking about the universe and the uniqueness of intelligence is one of the reasons I wanted to study the brain. But there are plenty of other reasons right here on Earth. For example, understanding how the brain works has implications for medicine and mental health. Solving the brain's mysteries will lead to true machine intelligence, which will benefit all aspects of society in the same way that computers did, and it will lead to better methods of teaching our children. But ultimately, it comes back to our unique intelligence. We are the most intelligent species. If we want to understand who we are, then we have to understand how the brain creates intelligence. Reverse engineering the brain and understanding intelligence is, in my opinion, the most important scientific quest humans will ever undertake.

When I started on this quest, I had a limited understanding of what the neocortex did. I and other neuroscientists had some notions about the brain learning a model of the world, but our notions were vague. We didn't know what such a model would look like or how neurons could create it. We were awash in experimental data and it was difficult to make sense of the data without a theoretical framework.

Since that time, neuroscientists around the world have made significant progress. This book focuses on what my team has learned. Much of it has been surprising, like the revelation that the neocortex doesn't contain one model of the world but about 150,000 sensory-motor modeling systems. Or the discovery that everything the neocortex does is based on reference frames.

In the first part of this book, I described the new theory of how the neocortex works and how it learns a model of the world. We call this the Thousand Brains Theory of Intelligence. I hope my exposition was clear and that you found my arguments compelling. At one point I debated whether I should end right there. A framework for understanding the neocortex is certainly ambitious enough for one book. However, understanding the brain naturally leads to other issues of consequence, so I kept going.

In Part 2, I argued that today's AI is not intelligent. True intelligence requires machines to learn a model of the world the same way the neocortex does. And I made the case for why machine intelligence is not an existential risk, as many others believe. Machine intelligence will be one of the most beneficial technologies we will ever create. Like every other technology, there will be people who abuse it. I worry more about that than AI itself. On its own, machine intelligence does not represent an existential risk, and the benefits, I believe, will be far greater than the downsides.

Finally, in Part 3 of the book, I looked at the human condition through the lens of intelligence and brain theory. As you can probably tell, I am concerned about the future. I am concerned about the welfare of human society and even the long-term survival of our species. One of my goals is to raise awareness of how the combination of the old brain and false beliefs is a real existential risk, far greater than the presumed threat of AI. I discussed different ways we might reduce the risks we face. Several of them require that we create intelligent machines.

I wrote this book to convey what my colleagues and I have learned about intelligence and the brain. But beyond sharing this information, I hope to convince some of you to act on it. If you are young or contemplating a career change, consider entering the fields of neuroscience and machine intelligence. There are few subjects more interesting, more challenging, and more important. However, I must warn you: it will be difficult if you want to pursue the ideas that I wrote about in this book. Both neuroscience and machine learning are large fields with tremendous inertia. I have little doubt that the principles I described here will play central roles in both areas of research, but it might take years to happen. In the meantime, you will have to be determined and resourceful.

I have one more ask, which applies to everyone. I hope that one day every person on Earth will learn how their brain works. To me, this should be an expectation, like, "Oh, you have a brain? Here is what you need to know about it." The list of things everyone should know is short. I would include how the brain is composed of the new part and the older parts. I would include how the neocortex learns a model of the world, whereas the older parts of the brain generate our emotions and more primitive behaviors. I would include how the old brain can take control, causing us to act in ways we know we shouldn't. And I would include how all of us are susceptible to false beliefs and how some beliefs are viral.

I believe everyone should know these things, in the same way that everyone should know that the Earth orbits the Sun, and that DNA molecules encode our genes, and that dinosaurs lived on Earth for millions of years but are now extinct. This is important. Many of the problems we face—from wars to climate change—are created by false beliefs or the selfish desires of the old brain or both. If every human understood what was going on in their head, I believe we would have fewer conflicts and a sunnier prognosis for our future.

Each of us can contribute to this effort. If you are a parent, teach your children about brains in the same way you might hold up an orange and apple to teach your children about the solar system. If you write children's books, consider writing about the brain and beliefs. If you are an educator, ask how brain theory can be included as part of a core curriculum. Many communities now teach genetics and DNA technologies as a standard part of their high school curricula. I believe brain theory is equally if not more important.

What are we?
How did we get here?
What is our destiny?

For millennia, our ancestors have asked these fundamental questions. This is natural. We wake up and find ourselves in a complex and mysterious world. There is no instruction manual to life and no history or backstory to explain what it is all about. We do our best to make sense of our situation, but for most of human history, we have been ignorant. Beginning a few hundred years ago, we started answering some of these fundamental questions. We now understand the chemistry underlying all living things. We understand the evolutionary processes that led to our species.

And we know that our species will continue to evolve and will likely become extinct sometime in the future.

Similar questions can be asked about us as mental beings.

What makes us intelligent and self-aware?
How did our species become intelligent?
What is the destiny of intelligence and knowledge?

I hope I have convinced you that not only are these questions answerable, but we are making excellent progress in answering them. I hope I have also convinced you that we should be concerned about the future of intelligence and knowledge, independent of our concern about the future of our species. Our superior intelligence is unique, and as far as we know, the human brain is the only thing in the universe that knows the broader universe exists. It is the only thing that knows the size of the universe, its age, and the laws by which it operates. This makes our intelligence and knowledge worthy of preservation. And it gives us hope that one day we may understand everything.

We are Homo sapiens, the wise humans. Hopefully, we will be wise enough to recognize how special we are, wise enough to make the choices that ensure our species survives as long as possible here on Earth, and wise enough to make the choices that ensure that intelligence and knowledge survive even longer, here on Earth and throughout the universe.

Suggested Readings

People who have heard about our work often ask me what I recommend reading to learn more about the Thousand Brains Theory and the related neuroscience. This usually prompts a deep sigh from me, because there isn't a simple answer, and to be honest, it is hard to read neuroscience papers. Before I give you specific reading recommendations, I have some general suggestions.

Neuroscience is such a large field of study that even if you are a scientist intimately familiar with one subfield you might have trouble reading the literature in a different one. And if you are completely new to neuroscience it can be difficult to get started.

If you want to learn about a specific topic—say, cortical columns or grid cells—and you are not already fluent in that topic, then I recommend starting with a source such as Wikipedia. Wikipedia usually has multiple articles on any topic, and you can quickly jump between them by following links. It is the quickest way I know to get a feel for terminology, ideas, themes, etc. You will often find that different articles disagree or use different terminology. You will find similar disagreements in peer-reviewed scientific papers. As a rule, you need to read multiple sources to get a sense of what is known about a topic.

To dig deeper, the next thing I recommend are review articles. Review articles appear in peer-reviewed academic journals, but, as their name implies, they present an overview of a topic, including areas where scientists disagree. Review articles are usually easier to read than typical papers. The citations are also valuable because they present most of the important papers related to a topic in one list. A good way to find review articles is to use a search engine such as Google Scholar and type in something like "review article for grid cells."

Only after you have learned the nomenclature, history, and concepts of a topic would I recommend reading individual scientific papers. The title and abstract of a paper are rarely sufficient to know if it has the information you are looking for. I typically read the abstract. Then I scan the images, which in a well-written paper should tell the same story as the text. Then I jump to the discussion section at the end. This section is often the only place where the authors plainly describe what the paper is about. Only after these preliminary steps will I consider reading the paper from beginning to end.

Below are suggested readings by topic. There are hundreds to thousands of papers on each subject, so I can only give you a few suggestions to help you get started.

Cortical Columns

The Thousand Brains Theory is built upon Vernon Mountcastle's proposal that cortical columns have similar architectures and perform similar functions. The first reference below is Mountcastle's original essay where he proposed the idea of a common cortical algorithm. The second reference is a more recent paper by Mountcastle in which he lists numerous experimental findings that support his proposal. The third reference, by Buxhoeveden and Casanova, is a relatively easy to read review. Although it is mostly about minicolumns, it discusses various arguments and evidence related to Mountcastle's claim. The fourth reference, by Thomson and Lamy, is a review article on cortical anatomy. It is a thorough review of cellular layers and the prototypical connections between them. It is complicated, but it is one of my favorite papers.

Mountcastle, Vernon. "An Organizing Principle for Cerebral Function: The Unit Model and the Distributed System." In *The Mindful Brain*, edited by Gerald M. Edelman and Vernon B. Mountcastle, 7–50. Cambridge, MA: MIT Press, 1978.

Mountcastle, Vernon. "The Columnar Organization of the Neocortex." *Brain* 120 (1997): 701–722.

Buxhoeveden, Daniel P., and Manuel F. Casanova. "The Minicolumn Hypothesis in Neuroscience." *Brain* 125, no. 5 (May 2002): 935–951.

Thomson, Alex M., and Christophe Lamy. "Functional Maps of Neocortical Local Circuitry." *Frontiers in Neuroscience* 1 (October 2007): 19–42.

Cortical Hierarchy

The first paper below, by Felleman and Van Essen, is the one I mentioned in Chapter 1, which first described the hierarchy of regions in the macaque neocortex. I include it mostly for its historical interest. Unfortunately, it is not open-access.

The second reference, by Hilgetag and Goulas, is a more current look at issues of hierarchy in the neocortex. The authors list various problems of interpreting the neocortex as a strict hierarchy.

The third reference, a paper by Murray Sherman and Ray Guillery, argues that the primary way two cortical regions talk to each other is through a part of the brain called the thalamus. Figure 3 in the paper nicely illustrates this idea. Sherman and Guillery's proposal is often ignored by other neuroscientists. For example, neither of the first two references mention the connections through the thalamus. Although I did not talk about the thalamus in this book, it is so intimately connected to the neocortex that I consider it an extension of the neocortex. My colleagues and I discuss a possible explanation of the thalamic pathway in our 2019 "Frameworks" paper, which is discussed below.

Felleman, Daniel J., and David C. Van Essen. "Distributed Hierarchical Processing in the Primate Cerebral Cortex." *Cerebral Cortex* 1, no. 1 (January–February 1991): 1.

Hilgetag, Claus C., and Alexandros Goulas. "'Hierarchy' in the Organization of Brain Networks." *Philosophical Transactions of the Royal Society B: Biological Sciences* 375, no. 1796 (April 2020).

Sherman, S. Murray, and R. W. Guillery. "Distinct Functions for Direct and Transthalamic Corticocortical Connections." *Journal of Neurophysiology* 106, no. 3 (September 2011): 1068–1077.

What and Where Pathways

In Chapter 6, I described how cortical columns based on reference frames can be applied to what and where pathways in the neocortex. The first paper, by Ungerleider and Haxby, is one of the original papers on this topic. The second paper, by Goodale and Milner, is a more modern description. In it, they argue that a better description of what and where pathways is "perception" and "action." This paper is not open-access. The third paper, by Rauschecker, is perhaps the easiest to read.

Ungerleider, Leslie G., and James V. Haxby. "'What' and 'Where' in the Human Brain." *Current Opinion in Neurobiology* 4 (1994): 157–165.

Goodale, Melvyn A., and A. David Milner. "Two Visual Pathways—Where Have They Taken Us and Where Will They Lead in Future?" *Cortex* 98 (January 2018): 283–292.

Rauschecker, Josef P. "Where, When, and How: Are They All Sensorimotor? Towards a Unified View of the Dorsal Pathway in Vision and Audition." *Cortex* 98 (January 2018): 262–268.

Dendritic Spikes

In Chapter 4, I discussed our theory that neurons in the neocortex make predictions using dendritic spikes. Here I list three review papers that discuss this topic. The first, by London and Häusser, is perhaps the easiest to read. The second, by Antic et al., is more directly relevant to our theory, as is the third reference, by Major, Larkum, and Schiller.

London, Michael, and Michael Häusser. "Dendritic Computation." *Annual Review of Neuroscience* 28, no. 1 (July 2005): 503–532.

Antic, Srdjan D., Wen-Liang Zhou, Anna R. Moore, Shaina M. Short, and Katerina D. Ikonomu. "The Decade of the Dendritic NMDA Spike." *Journal of Neuroscience Research* 88 (November 2010): 2991–3001.

Major, Guy, Matthew E. Larkum, and Jackie Schiller. "Active Properties of Neocortical Pyramidal Neuron Dendrites." *Annual Review of Neuroscience* 36 (July 2013): 1–24.

Grid Cells and Place Cells

A key part of the Thousand Brains Theory is that every cortical column learns models of the world using reference frames. We propose that the neocortex does this using mechanisms that are similar to what is used by grid cells and place cells in the entorhinal cortex and hippocampus. To get an excellent overview of place cells and grid cells, I recommend reading or listening to the O'Keefe and the Mosers' Nobel lectures, in the order they gave them. The three of them worked together to give a coordinated set of lectures.

O'Keefe, John. "Spatial Cells in the Hippocampal Formation." Nobel Lecture. Filmed December 7, 2014, at Aula Medica, Karolinska Institutet, Stockholm. Video, 45:17. www.nobelprize.org/prizes/medicine/2014/okeefe/lecture/.

Moser, Edvard I. "Grid Cells and the Enthorinal Map of Space." Nobel Lecture. Filmed December 7, 2014, at Aula Medica, Karolinska Institutet, Stockholm. Video, 49:23. www.nobelprize.org/prizes/medicine/2014/edvard-moser/lecture/.

Moser, May-Britt. "Grid Cells, Place Cells and Memory." Nobel Lecture. Filmed December 7, 2014, at Aula Medica, Karolinska Institutet, Stockholm. Video, 49:48. www.nobelprize.org/prizes/medicine/2014/may-britt-moser/lecture/.

Grid Cells in the Neocortex

We are only beginning to see evidence of grid-cell mechanisms in the neocortex. In Chapter 6, I described two fMRI experiments that showed evidence of grid cells in humans performing cognitive tasks. The first two papers—by Doeller, Barry, and Burgess, and Constantinescu, O'Reilly, and Behrens—describe these experiments. The third paper, by Jacobs et al., describes similar results from humans undergoing open-brain surgery.

Doeller, Christian F., Caswell Barry, and Neil Burgess. "Evidence for Grid Cells in a Human Memory Network." *Nature* 463, no. 7281 (February 2010): 657–661.

Constantinescu, Alexandra O., Jill X. O'Reilly, and Timothy E. J. Behrens. "Organizing Conceptual Knowledge in Humans with a Gridlike Code." *Science* 352, no. 6292 (June 2016): 1464–1468.

Jacobs, Joshua, Christoph T. Weidemann, Jonathan F. Miller, Alec Solway, John F. Burke, Xue-Xin Wei, Nanthia Suthana, Michael R. Sperling, Ashwini D. Sharan, Itzhak Fried, and Michael J. Kahana. "Direct Recordings of Grid-Like Neuronal Activity in Human Spatial Navigation." *Nature Neuroscience* 16, no. 9 (September 2013): 1188–1190.

Numenta's Papers on the Thousand Brains Theory

This book provides a high-level description of the Thousand Brains Theory, but it does not go into many details. If you are interested in finding out more, you can read my lab's peer-reviewed papers. They contain detailed descriptions of specific components, often including simulations and source code. All our papers are open-access. Here are the most relevant ones, with a brief description of each.

The following is our most recent paper, and also the easiest to read. It is the best place to start if you want to get a more in-depth description of the full theory and some of its implications.

Hawkins, Jeff, Marcus Lewis, Mirko Klukas, Scott Purdy, and Subutai Ahmad. "A Framework for Intelligence and Cortical Function Based on Grid Cells in the Neocortex." *Frontiers in Neural Circuits* 12 (January 2019): 121.

This next paper introduced our proposal that most dendritic spikes act as predictions, and that 90 percent of the synapses on pyramidal neurons are dedicated to recognizing contexts for predictions. The paper also described how a layer of neurons organized into minicolumns creates a predictive sequence memory. The paper explains many aspects of biological neurons that cannot be explained by other theories. It is a detailed paper that includes simulations, a mathematical description of our algorithm, and a pointer to source code.

Hawkins, Jeff, and Subutai Ahmad. "Why Neurons Have Thousands of Synapses, a Theory of Sequence Memory in Neocortex." *Frontiers in Neural Circuits* 10, no. 23 (March 2016): 1–13.

Next is the paper where we first introduced the idea that every cortical column can learn models of entire objects. This paper also introduced the concept of column voting. The mechanisms in this paper are extensions of the predictive mechanisms introduced in our 2016 paper. We also speculate that grid-cell representations might form the basis for the location signal, though we had not yet worked through any details. The paper includes simulations, capacity calculations, and a mathematical description of our algorithm.

Hawkins, Jeff, Subutai Ahmad, and Yuwei Cui. "A Theory of How Columns in the Neocortex Enable Learning the Structure of the World." *Frontiers in Neural Circuits* 11 (October 2017): 81.

The following paper extends our 2017 paper by working through in detail how grid cells can form a representation of location. It explains how such locations can predict upcoming sensory input. The paper proposes a mapping between the model and three of the six layers in the neocortex. The paper includes simulations, capacity calculations, and a mathematical description of our algorithm.

Lewis, Marcus, Scott Purdy, Subutai Ahmad, and Jeff Hawkins. "Locations in the Neocortex: A Theory of Sensorimotor Object Recognition Using Cortical Grid Cells." *Frontiers in Neural Circuits* 13 (April 2019): 22.

Acknowledgments

Though my name appears as the author, this book and the Thousand Brains Theory were created by many people. I want to tell you who they are and the roles they played.

The Thousand Brains Theory

Since its inception, over one hundred employees, post-docs, interns, and visiting scientists have worked at Numenta. Everyone contributed in one way or another to the research and papers we produced. If you are part of this group, I thank you.

There are a few people who deserve special mention. Dr. Subutai Ahmad has been my scientific partner for fifteen years. In addition to managing our research team, he contributes to our theories, creates simulations, and derives most of the mathematics underlying our work. The advances made at Numenta would not have happened if it were not for Subutai. Marcus Lewis also made important contributions to the theory. Marcus would often take on a difficult scientific task and come up with surprising ideas and deep insights. Luiz Scheinkman is an incredibly talented software engineer. He was a key contributor to everything we did. Scott Purdy and Dr. Yuwei Cui also made significant contributions to the theory and simulations.

Teri Fry and I have worked together at both the Redwood Neuroscience Institute and Numenta. Teri expertly manages our office and everything required to keep a scientific enterprise running. Matt Taylor managed our online community and was an advocate for open science and science education. He advanced our science in surprising ways. For example, he pushed us to livestream our internal research meetings, which, as far as I know, was a first. Access to scientific research should be free. I would like to thank SciHub.org, an organization that provides access to published research for those who cannot afford it.

Donna Dubinsky is neither a scientist nor an engineer, but her contribution is unsurpassed. We have worked together for almost thirty years. Donna was the CEO of Palm, the CEO of Handspring, the chair of the Redwood Neuroscience Institute, and is now the CEO of Numenta. When Donna and I first met, I was trying to convince her to take the CEO role at Palm. Before she had made up her mind, I told her that my ultimate passion was brain theory, and that Palm was a means to that end. Therefore, in a few years, I would look for a time to leave Palm. Other people might have walked out at that moment or insisted that I commit to Palm for the indefinite future. But Donna made my mission part of her mission. When she was running Palm, she would often tell our employees that we needed the company to be successful so that I would be able to pursue my passion for brain theory. It is not a stretch to say that none of the successes we had in mobile computing, and none of the scientific advances we made at Numenta, would have happened if Donna had not embraced my neuroscience mission on the first day we met.

The Book

It took eighteen months to write this book. Every day I would get to the office at around 7 a.m. and write until about 10 a.m. Although writing itself is solitary work, I had a companion and coach throughout, Christy Maver, our VP of marketing. Although she had no prior experience in writing a book, she learned on the job and became indispensable. She developed a skill for seeing where I needed to say less and where I needed to say more. She helped me organize the writing process and led review sessions of the book with our employees. Although I wrote the book, her presence is throughout. Eric Henney, my editor at Basic Books, and

Elizabeth Dana, the copy editor, made numerous suggestions that improved the clarity and readability of the book. James Levine is my literary agent. I can't recommend him more highly.

I want to thank Dr. Richard Dawkins for his delightful and generous foreword. His insights into genes and memes had a profound effect on my worldview, for which I am grateful. If I could pick one person to write the foreword, it would be him. I am honored that he did.

Janet Strauss, my spouse, read the chapters as I wrote them. I made several structural changes based on her suggestions. But more importantly, she has been the perfect partner for my journey through life. Together we decided to propagate our genes. The result, daughters Kate and Anne, have made our short stay in this world delightful beyond words.

Illustration Credits

Index

Jeff Hawkins is the cofounder of Numenta, a neuroscience research company, founder of the Redwood Neuroscience Institute, and one of the founders of the field of handheld computing. He is a member of the National Academy of Engineering and author of *On Intelligence*.